What Is Ecological Civilization?

Philip Clayton
Wm. Andrew Schwartz

What Is Ecological Civilization?

Crisis, Hope, and the Future of the Planet

In partnership with John B. Cobb, Jr, John Becker,
Pablo Diaz, Jack Patrick Sargent, & the
Institute for Ecological Civilization Team

ANOKA, MINNESOTA 2019

What Is Ecological Civilization? Crisis, Hope, and Future of the Planet

© 2019 Process Century Press

All rights reserved. Except for brief quotations in critical publications and reviews, no part of this book may be reproduced in any manner without prior permission from the publisher.

Process Century Press
RiverHouse LLC
802 River Lane
Anoka, MN 55303

Process Century Press books are published in association with the International Process Network.

Cover: Susanna Mennicke

VOLUME XVIII:
TOWARD ECOLOGICAL CIVILIZATION SERIES
JEANYNE B. SLETTOM, GENERAL EDITOR

ISBN 978-1-940447-41-4

Printed in the United States of America

SERIES PREFACE: TOWARD ECOLOGICAL CIVILIZATION

We live in the ending of an age. But the ending of the modern period differs from the ending of previous periods, such as the classical or the medieval. The amazing achievements of modernity make it possible, even likely, that its end will also be the end of civilization, of many species, or even of the human species. At the same time, we are living in an age of new beginnings that give promise of an ecological civilization. Its emergence is marked by a growing sense of urgency and deepening awareness that the changes must go to the roots of what has led to the current threat of catastrophe.

In June 2015, the 10th Whitehead International Conference was held in Claremont, CA. Called "Seizing an Alternative: Toward an Ecological Civilization," it claimed an organic, relational, integrated, nondual, and processive conceptuality is needed, and that Alfred North Whitehead provides this in a remarkably comprehensive and rigorous way. We proposed that he could be "the philosopher of ecological civilization." With the help of those who have come to an ecological vision in other ways, the conference explored this Whiteheadian alternative, showing how it can provide the shared vision so urgently needed.

The judgment underlying this effort is that contemporary research and scholarship is still enthralled by the 17th century view of nature articulated by Descartes and reinforced by Kant. Without freeing our minds of this objectifying and reductive understanding of the world, we are not likely to direct our actions wisely in response to the crisis to which this tradition has led us. Given the ambitious goal of replacing now dominant patterns of thought with one that would redirect us toward ecological civilization, clearly more is needed than a single conference. Fortunately, a larger platform is developing that includes the conference and looks beyond it. It is named Pando Populus (pandopopulous.com) in honor of the world's largest and oldest organism, an aspen grove.

As a continuation of the conference, and in support of the larger initiative of Pando Populus, we are publishing this series, appropriately named "Toward Ecological Civilization."

-John B. Cobb, Jr.

TABLE OF CONTENTS

Welcome Message, *i*

Preface 1
What Is Ecological Civilization?
John B. Cobb, Jr. 1

Introduction
Why These Particular Questions? 9

Question 1
Why "Civilization" and why "Ecological"? 15

Question 2
What Are the Underlying Causes of Ecological Catastrophe? 33

Question 3
Is "Ecological Civilization" Merely a Utopian Ideal? 45

Question 4
What Are the Foundational Insights of the Ecological Civilization Movement? 53

Question 5
What Other Movements are Allied with Ecological Civilization? 71

Question 6
How Does One Begin Building an Ecological Civilization? 97

Question 7
What Does Ecological Civilization Look Like in Practice? 113

Question 8
Why Does Ecological Civilization Bring Hope? 147

WELCOME MESSAGE

THIS BOOK is about the most urgent threat that humanity faces today, which may just make it the most important book you have ever held in your hands. A growing number of activists and scholars are framing the challenges of this threat—and the solutions—using the concept of ecological civilization. Unfortunately, these are not words that one hears all the time; for some, this book is actually the first time they will have heard the term at all. Therefore, we have attempted to be as straightforward and as non-technical as possible.

Hopelessness may just be our worst enemy at this point. However, the authors of this book, and the scholars and activists we have consulted, have discovered a fascinating feature of ecological civilization: the term brings hope unlike any other term we are aware of. For one, it helps us lift our eyes above the current threats to what will come afterwards. And, as it turns out, the idea is not just utopian; it actually offers guidance for lifestyle choices and policy formation in the present.

But there is another, and perhaps more surprising, strength to the term. It offers long-term hope, reminding us that humans have gone through many civilizations in the past. The end of a particular civilization does not necessarily mean the end of humanity, much less the end of all life on the planet. It is not hard for us to conceive of a society after the fall of modernity, in which humans live in an equitable and sustainable way with one another and the planet.

While ecological civilization offers long-term hope, it also provides values to guide the things we do today. You support co-ops and alternative economic models. You begin to farm the land around your house or in your city. You implement new forms of value-based education. You join an eco-village and seek to lower your impact on the planet to one "carbon footprint." As you make these changes, you begin to realize that these are concrete steps toward the development of an ecological civilization. Your actions, and the actions of more and more people around you, are actually part of birthing a new kind of civilization. We don't yet know whether these acts will avoid the economic and social collapse that threatens the planet. But even if they don't, they are still setting the course for what comes afterwards.

This volume is the result of hard work from several people who helped research and formulate some of the material, which was then compiled and used by the two authors. It is a team-written book, and no one person gets all the credit. Contributing to the result were John Becker, John Cobb, Pablo Diaz, Marilyn Greenberg, Jack Patrick Sargent, and the Institute for Ecological Civilization team as a whole. We owe a special debt of gratitude to our mentor, John Cobb, whose vision of ecological civilization has strongly influenced all of us. But we also owe much to other mentors, scholars, scientists, political theorists, economists, and visionaries whose work has advanced and directed our thought. We are reminded that the vision of "ecological civilization" comes to us from Chinese scholars and policy makers, from whom we have learned much.

What these pages offer to you is far richer because of the work of the 1500 members of the ecological civilization working groups at the 2015 "Seizing an Alternative" conference at Pomona College in Claremont, California, and the support of some additional 500 people who came to the keynote addresses. We express our gratitude to all those people, known and unknown, who have thought and acted prophetically to begin to move humanity in a new direction.

PREFACE

WHAT IS
ECOLOGICAL CIVILIZATION?

John B. Cobb, Jr.

THE TERM "ecological civilization" comes close to being an oxymoron, that is, an internally self-contradictory term, like "a black-haired blond." That our goal is threatened by that status indicates the extreme difficulty of the task that lies before us. "Ecological" points to the natural world that develops the complexes of living things that are possible in all sorts of contexts. We humans are impressed by the richness of the systems that nature develops and by their resilience in the face of changes in climate. We recognize that human involvement in an ecosystem almost always impoverishes it. Perhaps this is not true of some gathering and hunting societies, but it is sometimes true even of them. And it has certainly been true of all civilizations.

Indeed, "civilization" is partially defined in terms of humans altering their environment in favor of the immediate desires of their species. At least some gathering and hunting societies do little of that, instead adapting themselves to the environment. In contrast, a civilization changes the environment intentionally and successfully, so as to provide what humans want. In this sense, it might seem, a civilization is inherently anti-ecological. How then can we describe the goal for humanity as transitioning to ecological civilization?

WHAT IS ECOLOGICAL CIVILIZATION?

To achieve an ecological civilization would not be to give up modifying nature, but to learn to do so in ways that we can learn from nature and from its success in creating ecosystems that over time increase in complexity and richness. In general, natural ecosystems enrich the soil; that is, they multiply the life in the soil, which then, in turn, increases the quality of the ecosystem it supports. Unfortunately, most human civilizations have been accompanied by forms of agriculture that involve replacing natural ecosystems with monocultures. The grains that have played the central role in feeding growing populations have required annual ploughing, which always involves some loss of topsoil. As the human population has grown over the last few hundred years, and agriculture has been transformed into agribusiness, the loss of topsoil has accelerated. The inherently unecological character of modern civilization is now visible on a global scale.

And yet viable alternatives exist, ready to be implemented. Consider the work of Wes Jackson of the Land Institute in Kansas, who demonstrated that farmers do not have to raise food using only annual monocultures. Over millennia, monocultures have been the easiest way to increase food production. But when a practice proves unsustainable, a human society can learn from nature how to produce food in ways that regenerate the land. What Jackson learned from the prairie was that it could produce abundantly in ways that deepened and enriched the soil. It did so with a polyculture of perennials instead of the monoculture of annuals that agribusiness has imposed on farmland around the world.

Humans had imposed their monoculture of annuals so as to produce more of what humans directly require—grain seeds. Most agriculturalists insisted that there was no other option. To move to perennials, they argued, would be to accept great reductions in production at a time when food was beginning to be globally scarce. But Jackson did not agree, and over several decades he developed one grain, kernza, that is perennial and yet produces as much humanly edible seed as any annuals. Other grains are on the way, and there is still much to be done to shift from monocultures to the sort of polycultures that nature has shown to be healthiest. But that this is impossible has now, once and for all, been refuted. One enormous obstacle to a global transformation of food production has been overcome. Similar solutions exist for the transition away from a meat-based diet, which has massively damaging environmental consequences.

Humanity *can* feed itself ecologically. Of course, this will require enormous changes in customs and practices. It will not occur by itself. Shifting to ecological food production will not be easy. One success is not enough. But there is now good reason to believe that the idea of ecological civilization is *not* an oxymoron. Civilized people can learn from nature how to create a sustainable global society.

Some years ago, Paolo Soleri proposed that, instead of continuing to build unecological cities, we should build architectural ecologies, which he called "arcologies." Sadly, none has been built. We have continued to build in a way that removes thousands of square miles of fertile land from food production and requires an enormous amount of energy for heating and cooling and transportation. How different this would all be if we had paid attention to Soleri fifty years ago.

Since there still seems to be little interest in building cities in an ecological way, one might ask why I bring this up at all. It is to show that whether we develop an ecological civilization is a matter of our choosing. There is nothing objectively impossible about the transition. Cities could easily meet all the needs they are designed to meet with a relatively small amount of solar energy. And they could occupy a tenth of the space they now cover. Far more land could be left for agriculture, recreation, and wilderness. Cities could fit into a context of natural ecologies and agricultural ecologies, and their internal life could also have an ecological character.

By an ecological life among human beings, I mean healthy communities. Humans are inherently social beings. If they find themselves in flourishing communities, life will be good, and they will enjoy contributing to the groups in which they find themselves. These local groups will be able to shape their own lives as they desire to a large extent. This would be easier in an arcology than in our sprawling suburbs and impersonal apartment buildings, but much can still be done to organize ourselves in neighborhoods in which people take some responsibility for one another and for the group as a whole.

Beginning from where we now are, these are difficult goals. But they are not impossible. They are about recovering something we humans once had and have too often lost. The idea of community life and practical self-government is not sheer fantasy. In some parts of the world, and it is in those where people are happiest, it exists even now.

The term "ecological" enriches the idea of community. Communities are easiest to develop among homogenous people. It would be a serious mistake to condemn homogeneous communities. But they are not the ultimate ideal and, in any case, they are likely to become rarer. An ecosystem is composed of a great variety of entities and especially of living things. Although we would not want all of the types of relations found in an ecosystem to be copied in a human community, we have learned that variety, even quite basic difference, can enrich a community rather than destroy it.

However, by itself, having numerous local communities does not constitute or insure an ecological civilization. Human beings have a strong tendency to organize their view of one another in terms of "us" and "them." The modern view of human beings as self-enclosed and self-interested individuals has little support in history. But there is lots of support for the view that human beings will give their lives, if need be, for their tribe, or nation, or some other "we." This is noble and endearing, but too often their devotion to their own community is expressed in hostility to others. Too often this results in mutual slaughter. What is locally healthy and fulfilling turns out to be perhaps the greatest obstacle to a global ecological civilization.

We cannot reduce our concern for local communities. An ecological civilization can include a few extreme individualists, but individualism cannot sustain a civilization as a whole, or any of its parts. An ecological civilization will be one in which each community considers itself to be in community with other communities. In a world in which our national propaganda demonizes other communities in order to make us willing to make sacrifices for our own, this alternative may seem unrealistic. But, in fact, it is a natural part of life for all of us.

Let us suppose that we live in a village that has a strong sense of identity. We will consider the other villagers as part of us and take pride in what our village achieves. But in most countries, we will also have a wider sense of "we" that includes not only our own village, but neighboring villages as well—perhaps a county. And taking pride in the county, even in competition with other counties, does not prevent us in other contexts from identifying strongly with our state. And, in fact, in our world, the sense of "us" is likely to be strongest at the national level. To call for a nation to be a community of communities of communities of communities is not a drastic change from what often already exists.

To be together with others in a community does not preclude competition. Team sports create a strong sense of community in competition with other teams, but only rarely does it create genuinely hostile or destructive feelings toward competitors. Some competition among neighborhoods may be incompatible with inclusive community, but other forms can be healthy. Neighborhoods can compete to see which can deal best with their trash or most reduce the use of fossil fuels. This will not prevent them from cooperating on other projects.

The most serious problem in our present global system is that at some point there is no larger "we." Americans may sometimes consider that the "we" includes friends and allies, but thus far this has always been over against "them": the dictators, the Communists, the terrorists, the "axis of evil." This is the greatest political challenge. Can we unite in facing the dangers to our species and to our planet so that we really think of "we humans," or even "we living beings?" Yes, it is possible. Millions of people already think this way. But thus far those who control our media and our education do not want this, and they are able to use the very deep tendency to feel threatened by "them" to gain their ends in our society.

Obviously, there is much more to be said about an ecological civilization. Civilization involves education, so **ecological civilization** requires ecological education (Finland is perhaps leading the way here). It requires an ecological economics, and that has been quite fully developed in theory, although a very unecological economics still dominates the world. It requires that technology and science rethink their self-understanding and role in the service of an ecological civilization.

WHAT IS ECOLOGICAL CIVILIZATION?

My hope has been to illustrate what is involved in an ecological civilization without claiming to spell out the one pattern that all peoples should adopt. Variety is important. The idea that everyone voting is the *sine qua non* of healthy politics should not, I think, be universalized, although to me it seems that in an ecological society people should have as much control as possible over their own lives and the rules under which they live. Am I going too far? Perhaps. It seems to me that in schools we should encourage group activities and accomplishments and not organize so much around individual competition. Am I becoming too specific?

Those of us who are committed to working toward an ecological civilization, at least at this stage, propose many specifics. Some may be simply rejected. Others may be affirmed for some cultures but not others. Some may hold up under criticism so as to become goals for all. We understand that global ecological civilization should contain a variety of regional and local ecological civilizations. To join in working toward ecological civilization is to join the conversation about what it would be like everywhere or in some one location.

I have indicated my strong judgment that its agriculture will be based on a polyculture of perennials. But if someone can show that this should apply only in limited areas and the approach to agriculture elsewhere should be different, the discussion of what constitutes an ecological civilization may be enriched. I expect, indeed, I hope, that a decade from now the ideas about how a civilization can be ecological will have developed so that this book will be recognized as an expression of one stage in the overall movement. Everyone who truly cares that humanity has a healthy future on this planet, and who recognizes that much that we are now doing works against that future, is welcome to join us. All are also free to work with us to find better language to talk about our goals.

In closing, I would like to add a word about the history of the idea. It originated in Russia, but was first fully developed in China. Eventually it won its way into the constitution of the Chinese Communist Party. I suspect that the lack of opposition came from the assumption of many that it was being considered as a "postmodern" goal. They may have been fully committed to modernization and thought that first China would modernize. Only then, in a rather remote future, would China go on to perfection.

Others, of course, understood it to mean a strong emphasis on ecological matters here and now. Some of them were involved with Zhihe Wang and Meijun Fan in promoting what they called "the second enlightenment," strongly influenced by the philosophy of Alfred North Whitehead. As a result, in China Whitehead's thought and ecological civilization are related in many people's minds. This connection was strengthened by the fact that the Tenth International Whitehead Conference in 2015 featured the goal of ecological civilization.

We in Claremont were invited to take the lead in annual conferences spelling out the meaning and practical implications of "ecological civilization." We have held twelve so far. In these conferences and in our work in China we have, with some success, made the goal of ecological civilization controversial by suggesting that it subordinates the aim to modernize.

The transition to organic, systemic, and ecological thinking is not dependent on a single philosopher, leader, or country. Many people are reaching the basic ideas independently. The chapters that follow include a variety ideas, voices, and approaches. One does not have to be a scholar or philosopher to understand them. But reflection on what the goals are, and what major transformations will be necessary for us to achieve them, is a crucial step. We hope this brief introduction will guide and inspire readers to embrace, and to work toward, a genuinely ecological civilization.

INTRODUCTION

WHY THESE PARTICULAR QUESTIONS?

John Cobb, perhaps the best-known proponent of ecological civilization in North America, has just offered a first introduction to our topic. Before we turn to the eight central questions about ecological civilization, it will be helpful to formulate the core assumptions on which our exploration is built:

- We face a twofold task: We need to understand what is meant by ecological civilization if it is to function as a meaningful goal. And we need to clarify what are its implications for action and policy across the various sectors of society.

- A number of conceptual sources contribute to the notion of ecological civilization. It has philosophical roots in a philosophy of organism and internal relations (instead of mechanism), such as one finds in the process philosophy of Alfred North Whitehead and related thinkers. It receives theoretical support from the ecological sciences, learns from connections with systems thinking and network theory, and draws inspiration from the world's religious and spiritual traditions.

- Working toward an ecological civilization is not a romantic or utopian ideal. It does not mean returning to a pristine primordial state where humans have no impact on nature. It does however require learning to live in harmony with nature, helping to create ecosystems that increase in complexity and richness over time.

- Underlying the language of ecological civilization is a holistic vision in which all aspects of society are rethought in terms of the ecological frame: agriculture, education, governance, economics, community, and more.

- Alongside its theoretical content, the goal of ecological civilization has many specific implications. It involves concrete changes to how we farm, how we build our cities, and how we achieve healthy communities.

- Humanity does have the capacity to make the transition beyond the destructive patterns of modern civilization. The longer we continue our current rates of consumption, the more difficult the transition becomes.

- Humans will only succeed at bringing about these changes if we make the transition from tribal thinking—"us" *versus* "them"—to "we" thinking. We must move beyond the modern view that human beings are like individual self-enclosed atoms, driven by self-interest above all else.

- An ecological civilization implies a global network of communities within communities within communities. The network extends downward to local communities, perhaps on the model of traditional village life; and it extends upward to a global awareness of our interdependence with each other, with other life forms, and with the planet as a whole.

The first step, then, is to attempt to give the notion of ecological civilization a clearer definition. We will examine its scientific grounding in the ecological sciences and its historical basis in the history of civilizations. We will be honest about its difficulties and appreciative of the similar terms and ideas used by our allies in related movements. We will trace the impact that the transition toward an ecological civilization is already having around the world. Finally, we will outline some of the steps that humanity must take as we move beyond the epoch known as modernity and toward the emergence of a new form of civilization.

Taken together, the eight specific questions that follow offer, we believe, the clearest and most direct way to understand what ecological civilization is. These are not the only questions that could have been discussed, nor do we wish to discourage additional questions from being considered. But each of the questions explored in this book were chosen for a particular purpose.

(1) *Why "Civilization" and Why "Ecological"?* In answering this question, we begin to unpack the notion of an ecological civilization as a unique forum of human community. What most clearly distinguishes an ecological civilization from other civilizations is the relation to the natural world. Whereas a defining characteristic of other civilizations is the manipulation of the environment for the sake of the people, an ecological civilization modifies the environment in sustainable and symbiotic ways, considering the well-being of nature as well as people. This is not a romantic ideal of a return to the pristine purity of nature; it represents, after all, a form of civilization. Not simply about humans living in harmony with nature, ecological civilization also entails humans living in peaceful ways with other humans, to promote the flourishing of all life.

(2) *What Are the Underlying Causes of Ecological Catastrophe?* In answering this question, two things happen. First, we begin to reorient our thinking toward understanding the underlying causes—not just the symptoms—of our world's most urgent threats, which is essential to adequately addressing these problems. Second, we recognize that the underlying causes of the ecological crisis are inseparable from social and economic crises. The massive inequality between the rich and the poor is not separate from an economics of unlimited growth and the depletion of natural resources, extinction of species, or global warming.

WHAT IS ECOLOGICAL CIVILIZATION?

Contemporary civilization is designed to benefit the privileged elite at the expense of the poor and the environment. We need a new civilization that promotes the common good—human and nonhuman alike.

(3) *Is "Ecological Civilization" Merely a Utopian Ideal?* Ecological civilization is not an unattainable ideal, but a vision for a better world that is already beginning to emerge. While utopian ideals portray carefree futures of peace and abundance, dystopian visions portray futures of conflict and scarcity. While utopias tend to depict cooperation and sharing, dystopias depict violent individualism where survival requires looking out for one's personal interests alone. Ecological civilization stands between utopias and dystopias by outlining the requirements for a future of material sufficiency and ecological balance, while recognizing that the struggle is real. The task is as much about clarifying, and taking, the steps toward an ecological civilization as it is about outlining the ideal itself. Also, it's an ongoing process and not simply an arrival point; the further humanity moves in building the various sectors of society on ecological principles, the more we will see further possibilities for improvement.

(4) *What Are the Foundational Insights of the Ecological Civilization Movement?* Answering this question takes us deeper into the history, philosophy, and fundamentals of the ecological civilization movement. We all have assumptions about the world, most of which go unexamined in day-to-day life. The vision of ecological civilization is enriched by those in the Seizing an Alternative movement who have given much thought to the fundamental philosophical issues at play. All things are interconnected, so we need systematic, comprehensive, and disciplined processes for mapping an alternative future. Since the problems are interrelated, the solutions can't be piecemeal. All life is intrinsically valuable, so we need a new story—a new paradigm—that appropriately conveys humanity's place on this planet. Ideas are important because practice is never free from theory. Whether we realize it or not, our systems and practices, both societal and personal, are often informed by unexamined assumptions about the world. Therefore, a change in one's world- and life-view should change one's life and practice. After

the paradigm shift in thinking come the concrete details.

(5) *What Other Movements Are Related to Ecological Civilization?* While an ecological civilization needs a foundation, it won't be built on a single stone. As we struggle to conceive radically different ways of living on the planet, diversity helps increase the complexity and richness of our conceptions. Therefore, it is important to explore parallel movements that, with their distinct formulations of the foundational concepts, offer similar visions for a better future. That is the function of question five. From integral ecology and the goal of an ecozoic era, to organic Marxism, the Earth Charter, and more, this section examines distinct movements with shared core values, philosophies, and hopes for the future—each calling for fundamental systemic change across all areas of society.

(6) *How Does One Begin Building an Ecological Civilization?* Answering this question begins the transition from theory to practice. By this stage, you'll have a basic understanding of the notion of ecological civilization—what it is, what it's not, and how it relates to similar movements. This chapter builds upon that foundation, moving from worldview-level concerns to their concrete implications for life and action. How do we begin to transform our forms of community, our systems of economics and production, our models of education, and the functions of culture, tradition, religion, and spirituality? This section takes us from the contemporary crisis to a clearer understanding of the steps required for systemic change.

(7) *What Does Ecological Civilization Look Like in Practice?* This may be one of the most important topics addressed in this book. To answer the question is to bring the vision of ecological civilization (the "big idea") down to earth in more tangible ways. Theory informs practices, which in turn inform theory. By looking at concrete examples of how ecological civilization is already being realized, the vision becomes that much clearer. And as the vision becomes clearer, so do the concrete steps for personal and societal improvement.

(8) *Why Does Ecological Civilization Bring Hope?* We are in an unprecedented period of human history. For the first time, humanity has the potential to destroy the very capacity of our planet to sustain

life. Yet it's also the first time we've had the ability to create a truly ecological civilization. While the scope of our global crisis, its environmental and social impacts, can produce a paralyzing despair, the realistic prospect of creating a world that works for all people and the environment can also impart hope. While hopelessness may be our greatest enemy, a well-founded hope may be humanity's greatest advantage. This closing section will explore the various ways in which ecological civilization can bring hope.

QUESTION 1

WHY "CIVILIZATION" AND WHY "ECOLOGICAL"?

THE MOST DIRECT WAY to think about the title of this book is to explore each of the two words separately—*civilization* and *ecological*—to better understand how the terms relate to one another.

What Is a Civilization?

Civilizations are particular ways that groups of humans—societies, cultures, countries—organize their lives together. What's unique about the term is that it expresses the *largest* or *broadest* framework that we can find for expressing commonalities across large groups of people, large regions of the world, and large expanses of time.

As John Cobb explained in his opening essay, the rise of civilization came with the movement away from hunting and gathering societies that were tightly adapted to their natural environments, to societies based on agricultural settlements, which began to alter the natural environment for the benefit of the people. Modifying nature is a distinctive mark of civilization. But the modification of nature on the global scale has led to the commodification of nature that lies at the heart of our current crisis. How we relate to our environment is a fundamental matter, but it is not separate from how we relate to one another.

WHAT IS ECOLOGICAL CIVILIZATION?

Groups of people structure their life together in quite basic ways: What do its members eat? How do they dress? How do they approach family? How do they express themselves in art and music? What is sacred for them, and what is profane? These distinctive patterns are their *culture*. Cultures can be as specific as the insider language of a group of Middle School friends, or as broad as the Greco-Roman world.

Recognizable patterns are often shared across diverse cultures. When these patterns include writing and social organization, and when they cover extensive regions and endure over extensive periods of time, we say that a group of cultures is bound together into a single civilization. In one sense, *a civilization is the sum total of the cultures that it encompasses*. It's a style, a set of practices and fundamental values that its members share. In another sense, a civilization is far more than the sum of its parts. Often, the fundamental values and practices of a civilization are tied to an underlying set of assumptions that pervade all that its members do—assumptions so deep that people may not even be aware of them.

The term "civilization" can be traced back to the Latin root *civilis*, which means "citizen" or "city." The term was not widely used until the 18th century, however. Most commonly, the term 'civilization' refers to a large collection of cultures that endures over time and expands over large regions. The term can also refer to societies that are considered advanced, sophisticated, or enlightened (i.e., civilized). Unfortunately, this notion of civilization has been used to support colonialism and the development of empires in which the "civilized" groups would conquer the uncivilized "barbarians" in order to save them from their ignorance. Needless to say, civilization in this sense has contributed to countless atrocities across human history.

In the context of an "ecological civilization," the term "civilization" refers broadly to *a way of living together with shared values*. It encompasses everything from agriculture and economics, to governance, education, religion, transportation, medicine, architecture, art, music, and others. It is this broader sense of civilization that is being used by 21st century scholars to identify and study vastly different civilizations around the world and across the ages. Civilization Studies has allowed us to recognize similarities and differences as never before.

Question #1

Modern Civilization

Our present civilization (the one for which an ecological civilization is the needed alternative) is commonly called "modern civilization," or modernity for short. The philosopher Charles Taylor is perhaps the leading contemporary scholar of modern civilization. He has shown how the dominant institutions of modernity have come to seem self-evident to people today, stating:

> My hypothesis is that central to Western modernity is a new conception of the moral order of society. At first this moral order was just an idea in the minds of some influential thinkers, but later it came to shape the social imaginary of large strata, and then eventually whole societies. *It has now become so self-evident to us, we have trouble seeing it as one possible conception among others.* The mutation of this view of moral order . . . is the development of certain social forms that characterize Western modernity: the market economy, the public sphere, the self-governing people, among others.[1]

Taylor emphasizes three distinctive features of modern civilization:

- we came to imagine society primarily as an economy for exchanging goods and services to promote mutual prosperity;
- we began to imagine the public sphere as a metaphorical place for deliberation and discussion among strangers on issues of mutual concern; and
- we invented the idea of a self-governing people capable of secular "founding" acts without recourse to transcendent principles.[2]

These practices and values are so obvious, so self-evident to moderns, that they have become invisible to us. For this reason, they appear unchangeable. But are they? Consider five of the conceptual foundations of modernity:

1. It's often said that modernity began around 1600. René

Descartes, the so-called father of modern thought, defended a turn to the subject as the sole source of meaning and value. Descartes viewed animals as mere machines and nature as a background for the adventures of mind and thought. He thus dichotomized the world into "thinking things" and "extended things." This move is not self-evident; most other civilizations have begun with groups or societies of persons-in-community.

2. Modern philosophy has been largely based on the model of competition and domination. Already in 1649 the political philosopher Thomas Hobbes affirmed that the human condition is "a war of all against all," so that life on earth is "nasty, brutish, and short." Charles Darwin's theory of natural selection was quickly interpreted by his peers as affirming a nature "red in tooth and claw" (Alfred Lord Tennyson). This attitude of domination ran across all aspects of human existence and thought, including the domination of men over women, of lighter skin color over darker skin color, of more powerful nations over smaller nations, of science over religion, of humans over nonhumans, and of the rich over the poor. Of course, many forms of partnership existed. But *symbiosis*, the idea that working together is the more fundamental strategy for success, was widely rejected.

3. Modern civilization developed a particularly strong form of individualism. Society exists for the sake of the individual; analogously, nations exist exclusively for the sake of their citizens (e.g., "America first"). In his *Second Treatise*, John Locke taught that the state exists for the sole purpose of preserving the life, liberty, and property of its (male) property-owning citizens. The philosopher John Stuart Mill became the spokesperson for this modernist ideal when he argued that the goal of the state is to maximize the freedom of each individual. John Stuart Mill then extended that tradition, insisting that the state should limit freedom only in cases where the actions of one person cause direct harm to another person. The actions of citizens belong in the "private sphere,"

Mill said, until they threaten harm to others; only directly harmful acts, like murder, belong in the public sphere and may be regulated by the state. Communitarianism—the view that individuals exist primarily for the good of society, the common good—rarely trumped individualism within the modern paradigm.

4. Scientific thinking was built on the foundation of mechanism. Descartes' "extended things" became for Thomas Hobbes the doctrine that all is "matter in motion," and for the 18th century French physician La Mettrie, man became a machine (*L'homme Machine*). Newtonian physics sought to explain all motion through laws of force acting on matter. "All life is chemistry," wrote Jan Baptist van Helmont in 1648. More recently, as we will see below, the biologist Richard Dawkins, the great critic of all things religious, sought to explain all human life and thought (i.e., civilization) in terms of the mastery of our "selfish genes" and, later, in terms of *memes*, the cultural analog to genes. Or, as one neuroscientist put it in a recent informal discussion, "wires and chemicals, that's all we are—wires and chemicals." The existence of moral principles and the possibility that religious beliefs might be true fall to the sword of scientific critique.

5. Finally, modernism has been, above all, an advocate of itself. All earlier phases of human history were "pre-scientific" and therefore inferior to the Scientific Age. Modern political philosophers argued that older political systems, philosophies, and forms of social and political organization have been superseded. The great progress that humanity has made is due to the principles of modernity, so that now the correct attitude is *meliorism*—the belief that society is getting better and better.

Examples of Other Civilizations

Of course our modern civilization isn't the only civilization. As John Cobb noted in the introduction, the emergence of civilization is often tied to the development of agriculture. The earliest civilizations emerged after 3000 BCE, when the rise of agriculture allowed people to have surplus food and increased security. Increased food security allowed a greater number of the population to look beyond basic survival toward other matters, including artistic and recreational activities. Hence, a civilization is often identified by the presence of large population centers, a written language, monumental architecture, unique art styles, systems for administering territories, a complex division of labor, and the division of the population into social classes. While civilizations emerged with agriculture, they would expand through trade, war, and exploration. Most civilizations would fall by either being incorporated into another expanding civilization, or by collapse and reversion to a simpler form.[3]

Civilizations first appeared in Mesopotamia (present-day Iraq), then in Egypt, the Indus Valley (2500 BCE), China (1500 BCE), and Central America (present-day Mexico, 1200 BCE). The list of major civilizations over the centuries is staggering. In addition to those already listed, there were the Mayan, Greek, Persian, Roman, Aztec, Inca, Elamite, Hurrian, Osirian, Zapotec, and Hattian civilizations (to name a few).[4] Given that civilizations have come and gone on every continent (except Antarctica) over the last 4,500 years, it is rather strange to think that modern civilization would not rise and fall as well.

African civilization has received the least attention. Eight thousand years ago, people in present-day Zaire developed their own numeration system, as did Yoruba people in what is now Nigeria. A structure known as the African Stonehenge in present-day Kenya (constructed around 300 BCE) served as a remarkably accurate calendar. The Dogon people of Mali amassed a wealth of detailed astronomical observations, including Saturn's rings, Jupiter's moons, the spiral structure of the Milky Way, and the orbit of the Sirius star system. African metallurgy included steam engines and carbon steel. Ancient Tanzanian furnaces could reach 1,800°C, up to 400°C warmer than those of the Romans.[5]

Mayan civilization was equally as advanced. It spread outward from the Yucatan Peninsula across Mexico and as far as Guatemala and Honduras. Mayan mathematicians discovered the zero, their astronomers predicted solar eclipses, their craftsmen produced rubber, and their engineers constructed a 100m suspension bridge and elaborate subterranean aqueducts where they could control water pressure.[6] Large cities allowed for significant cultural achievements in art, architecture, and written works. Many of these achievements were not matched in other civilizations for hundreds of years.

Civilization Studies

In centuries past, the breadth of a civilization was limited by mountains and oceans and distance. Thus the Hellenistic civilization could exist at the same time as the Chinese and Mayan civilizations. Today, for the first time, we exist in a single global civilization. The development of technologies for transportation and communication effectively erased the geographical challenges that once divided different civilizations. Cell phones, Hollywood movies, and the internet have done what armies and wars were never able to do; they have created a *global* community largely based on shared values, a balance (or imbalance) of power among nation states, and a system of global capitalism.

The University of Chicago has one of the world's strongest programs in Civilization Studies. Students have the chance to study the literature, art, architecture, politics, economics, and social organization of each major civilization of human history. In each case, professors seek to show how a civilization can offer an integrated and coherent framework for understanding the different ways that people organize their lives into societies. The University notes that its approach

> stresses the grounding of events and ideas in historical context and the interplay of events, institutions, ideas, and cultural expressions in social change . . . The courses emphasize texts rather than surveys as a way of getting at the ideas, cultural patterns, and social pressures that frame the understanding of events and institutions within a civilization. And *they seek to explore a civilization as an integrated entity, capable of developing and evolving meanings that inform the lives of its citizens.*[7]

Are civilizations good or bad? Well, both. They often begin by developing systems of agriculture so that more people can be nourished from the available land. Yet the centralized power that organizes these agricultural efforts often takes huge proportions for itself—forming a system of exploitation in which the poor and the land are overused. Similarly, civilizations often support the development of a shared language and writing system through which the arts and culture flourish and citizens enjoy experiences not available in a pre-civilizational context. And yet, again, the same mindset can also lead to a civilization becoming imperialistic and militaristic, bent on conquering and colonizing other lands, enslaving their population and destroying their culture. Both are true: the highest expressions of human culture have arisen in advanced civilizations; and yet competing groups have also tended to export or impose their own cultural values by force.

Civilizations end for a variety of reasons. When they grow old and weak, the populace may rise up in revolt, overturning their onetime leaders (think of the French Revolution in 1789, or the Russian revolutionaries who in 1918 killed the tsar, his relatives, and his family). Another common reason for a civilization to fall is that a competing civilization may conquer it. Along these lines, no power has been as successful at wiping out other civilizations as the Europeans and North Americans, authors of the world's first truly global civilization. Their colonialism has destroyed significant parts of the cultural heritage of the indigenous peoples of Africa, North America, South America, India, and China.

Today scientists and historians are also discovering how many civilizations ended through environmental collapse. Bloody revolts during the Ptolemaic Kingdom in ancient Egypt appear to have been caused by climate change, after volcanic eruptions blocked the monsoons and the lower water levels in the Nile led to food shortages.[8] (The data are worrisome because monsoons— which are still essential to water levels in the Nile and hence to food production—are increasingly being affected by climate change.) Other disasters are byproducts of overpopulation, which is similar to the explosion of the global population today. Often societies cut down the forests—as occurred on the Marshal Islands—or destroy the land that feeds them, turning rich soil into dead dirt so that crops can't grow. When people begin to starve, violence tends to follow. The heavy toll of large cities on the land around them has caused the collapse of numerous civilizations, as may have been the case with the Mayans.

Water pollution and water scarcity have played major roles. Lack of water purification in large cities, for example, led to plagues that could wipe out the population of an entire city. A new book by Kyle Harper, *The Fate of Rome: Calamity, Disease, and the End of an Empire*, gives evidence that *Yersinia pestis*, the bacterium that causes bubonic plague, was one of the major causes of the collapse of the Roman Empire. Bubonic plague or "the Black Death," caused by the same bacterium, killed some 25 million people between 1347 and 1352, roughly one-third of the population of Europe. Over the following decades, the resulting food shortages led to peasant revolts that helped bring about the end of medieval civilization.

Modernity and Globalization

Today people and countries around the world have bound themselves together into a single global civilization as never before. Their attraction to more and more powerful technologies has aligned consumers across the globe around a single set of desires and aspirations. As a result, the corporations who own and produce technological tools and toys have gained more and more massive wealth. They have helped mold the population of the planet into global citizens for the first time, which means that consumerism has succeeded where no army, no artistic work, and no cultural achievement could succeed. Presidents and dictators who fight against this tidal wave are being outvoted or overthrown by their citizens. The attractiveness of material success has become a global magnet, overcoming all resistance; global consumers won't take no for an answer. Smartphones—"the world in your pocket"—are one of many possessions that are now viewed as indispensable around the globe.

The myth of unlimited progress is seductive: medicine stopped plagues, and the death rate dropped. Today's problems will be solved by tomorrow's technology. We are only at the beginning of the comfort and efficiency that technology has in store for consumers. The move to greater prosperity and pleasure can continue indefinitely. Life will become more comfortable and enjoyable. People should own and consume whatever goods and possessions they desire; they need only to be wealthy enough to buy them. Thanks to global capitalism and the success of the markets, more wealth will continue to be generated, so people's lifestyles will continue to improve.

Most of this modern magic has been made possible by fossil fuels. Energy has been available to moderns as never before; most of the technology that we take for granted owes its existence to burning massive amounts of hydrocarbons. Online servers in the U.S. alone consume 70 billion kilowatt hours per year. It would take eight large nuclear reactors, or twice the output of all the nation's solar panels, to meet this energy demand.[9]

But now we have reached planetary limits. It's not that the world population would vote for a change of civilizations; after all, the modern one has made so many people happy—or at least it has made them long for the things that others have. Even a small farmer in India can now own a cell phone, and perhaps a TV. Residents of even the most isolated villages can wear Western T-shirts and drink Coca-Cola.

Unfortunately, the longer we live and the more we consume, the greater stress we place on the planet. Earth will give out before our desires do. The planet itself is bringing modern civilization to an end. Only now, in the late modern period, can we look back over the last few centuries and understand. After all, people only recognize in hindsight—often in a very late phase—what their civilization was.

The result from this brief study of civilization is clear: like all civilizations, it's inevitable that the modern civilization will end. It's not yet clear exactly when and how it will do so, but that we stand at the end of an age is certain. This raises the key question: could modernity be followed by a genuinely ecological, and thus a sustainable, civilization? Could humans find other ways to organize society and economics, and radically different practices in living with the land and other creatures—either to stave off a global collapse, or to rebuild civilization after the current one has run its course?

What Is "Ecological"?

The last few decades have seen explosive growth in the ecological sciences; *ecological* is now a household word.[10] The term covers both *facts* about how the living world is organized and *values* about how and why to preserve these natural ecosystems. Because our very existence depends on these ecosystems, in this case facts and values are inseparable for us. Ecosystems matter—they are valuable—because without them we could not survive as a species. Hence the facts about them are crucial to our life on this planet. As Holmes Rolston states,

> The term ecology is, etymologically, the logic of living creatures in their homes, a word suggestively related to "ecumenical," with common roots in the Greek *oikos*, the inhabited world.[11]

Advances in ecosystem studies, growing out of other scientific developments, have exploded many past myths about the world. It turns out that in fact not everything is reducible to physics, to "matter in motion" (Thomas Hobbes). Biological systems cannot be explained using the theories of physics alone. When the myth of reducing everything to physics collapses, so too does the myth of determinism. Physics may use deterministic equations (though that's only true of parts of physics), but by and large one can't predict biological evolution using them. Very small differences in the initial state of a biological system can produce vast (and unpredictable) differences down the line. You may recall the "butterfly effect": the realization that a small change in environmental conditions in the United States can have large, and perhaps devastating effects for the climate of northern Africa or Greenland.

Still, one last myth has to go: the myth that genes are the unchanging, determinative building blocks of biological systems. In the best-known defense of this philosophy, *The Selfish Gene*, Richard Dawkins maintained that genes are the central unit of biology and that they use the biological process to make replicas of themselves. Dawkins wrote, "Bodies are survival machines that genes construct to make more copies of themselves . . . We are . . . robot vehicles blindly programmed to preserve the selfish molecules known as genes."[12] He continues:

> I shall argue that the fundamental unit of selection, and therefore of self-interest, is not the species, nor the group,

nor even, strictly, the individual. It is the gene, the unit of heredity . . . They are in you and me; they created us, body and mind; and their preservation is the ultimate rationale for our existence. They have come a long way, those replicators. Now they go by the name of genes, and we are their survival machines.[13]

It's true that, when organisms reproduce, genes provide the blueprint. But they do not control all its future features.[14] Organisms adapt during the course of their lives, and not only through genetic mutations. The causes that influence organisms are not all "bottom up," with the genes specifying what they become. There are also important "top down" causes—changes that come from the environment and permeate downwards, from systemic impacts such as on the immune system and from there on down to impact the chemistry of individual cells. One of the more amazing discoveries is that the proteins in cells can actually change the genetic content itself, for example by transposing the rungs of the DNA "ladder." A new picture of organisms and the environment has grown from these discoveries.

The Heart of Ecology

Ecology is the science of interdependence *par excellence*. Of course, there are scientists who deny this, insisting that the ecological sciences are not really different from other sciences, since *all* science seeks to explain a set of complex interactions using more fundamental laws. This has meant that scientific success has been defined as traveling down a ladder of reductions, where the rungs are ecosystems, organisms, genes, biochemistry, chemistry, and, ultimately, fundamental physics. For example, there was a project at the Yale School of Forestry some time back that sought to calculate the total surface area of the leaves in a particular forest, and then to explain the forest's evolution using the biochemical processing capacities of the sum of all its leaves.

Most ecosystem studies, however, are not reductive in this way. Ecosystems are complex emergent realities that are more than the sum of their parts; the system is a complex integrated whole in terms of which the individual organisms are understood.[15] Reductionism fails because of this high degree of interdependence. For example, the thriving of one species is dependent on the reproduction rate and nutritional needs of others, and the complex balance between flora and fauna is necessary for both to survive. Very large and extremely small organisms engage in an intricate dance of interdependence. Their finely tuned symbiotic relationships represent a form of cooperation that increases their odds of survival.[16] Of course, there is still Darwinian competition: organisms better adapted to their environment outperform their competitors, and more of their offspring therefore reach reproductive age. But studies also show that inter-species cooperation also plays a crucial role in the survival and flourishing of ecosystems and the organisms of which they are composed.[17] Other organisms do not influence them externally only, but transform them internally as well. Waste products from a mammal or a fallen tree become nutrients for other species. Without understanding the rich interdependence of these systems, the behavior of their members cannot be understood.

Space does not allow for an exposition of similar patterns in all the life sciences, but suffice it to say that biology is full of such interdependencies. Our individual organs, such as liver, kidneys, and pancreas, are affected by our behaviors, and they in turn have holistic effects on the body. Our mental attitude, and emotional states, such as depression, affect the strength of our immune system and hence our susceptibility to disease. The influence is traceable all the way down to individual cells. Thought itself is influenced by small molecules knows as neurotransmitters, and they respond in turn to our moods and desires.

We have seen that the interdependence of ecosystems comes in part because their member organisms are not only externally related, but also internally related to each other. From genes to organisms to broader systems of cooperation, they mutually transform each other.

What Are the Core Principles of Ecology?

Ecology is an immensely complex field; a lifetime of work is required to comprehend even a single ecosystem or a single phase of evolution.

Nevertheless, the following is an attempt to summarize the field in just a few short paragraphs.

The starting point is *emergence*. At some point the conditions were right for the emergence of the first self-reproducing cell. This same principle of emergence describes the dynamic of all life. Every time organisms reproduce, minor variations arise. These can give rise to changes in the structure or behavior of the organism. Some of these structures make it better adapted to its environment. When that happens, it can create more little copies of itself, and they begin to fill the environment more quickly and effectively than others. Minor variations in one of their offspring lead to some that are better adapted . . . and so on, and so on.

What we see here is the growth in *complexity*. Over time more complex structures and behaviors emerge, which increase the survival advantages for organisms. What makes the study of biology so fascinating is that immensely complex systems arise through the simple interaction of these little variations with the surrounding organisms and environment.

It takes a lot of energy for this process to work; a planet must be close enough to its sun, but not so close that life is destroyed by the heat. When the conditions are right and evolution starts, no outside guidance is required. This is why we call organisms *self-organizing systems*: the complexity arises from the dynamics of evolution alone, as life-forms organize themselves in ever-different ways. It is an *open-ended process*. Unlike some processes in physics, where outcomes can be predicted over millions of years, the evolution of life is highly unpredictable, given the complexity of organisms and ecosystems and their finely-tuned interactions within others. To take just one example: the human brain has roughly 100,000,000,000 (10^{11}) neurons and 100,000,000,000,000 (10^{14}) neural connections. This makes it the most complex structure we have yet discovered in the universe. If we cannot fully predict the decay of even a single uranium atom, how could we ever predict the future responses of an entire human brain?

Amazing new structures and behaviors emerge over time. The result of this process is visible all around us: the polar bear's fur, the flamingo's color, the peacock's feathers, the gazelle's speed, chimpanzee communication, bonobo social structures, the dog-human relation, and the rise of a species that can create Harry Potter, recognize galaxies, and dream of world peace.

The wonder of life is visible in this interdependence. Even the simplest single-celled organism is the result of a complex, ongoing interaction between it and its surrounding environment. A relatively simple organism such as an earthworm interacts with its environment in millions of ways; the body and mind of a human being interact with the world billions of times each day. Obviously, we would not exist at all without the stable and nurturing life-systems around us. In fact, in one sense we are not separately existing creatures at all. We are, from the bottom up, *beings in community*. We are, from our simplest cell to our highest thought, organic expressions of the ecosystems that nourish us and give us life. We will have frequent opportunity to return to this organic thinking in the pages to come.

Conclusion: An Ecological Civilization

Until the 19th century, there had never been a global civilization—a single civilization that could overcome and eliminate all others. Civilizations flourished simultaneously in East Asia, South Asia (present-day India), Europe, Africa, and Latin America. When we think only of a single series of civilizations, the series that runs from Greece to the American Revolution, we imagine Western civilization as the norm and turn our backs on the rest.

While it lasted, each civilization made a deep impress on the lives of millions of people. Just as biodiversity is crucial for life on this planet, so also the rich history of our species never would have been possible without the cultural diversity spawned by this variety. Just naming the other civilizations and their legacies, as we have done in this chapter, is enough for one to see how ethnocentric this attitude is.

And yet, far more than we realize, civilizations come and go. Usually they last only a few hundred years. At close to 1,000 years, medieval European civilization is one of the major exceptions to the global pattern, and scholars question whether it should be seen as a single civilization at all. The vast majority of all human civilizations flourish for a few centuries, if that, and then go the way of all flesh. In the spirit of Shakespeare's *Macbeth*, we might say that each one is "a poor player / That struts and frets his hour upon the stage /And then is heard no more." Or, even more powerfully, remember Shelly's famous poem "Ozymandias," which describes the hubris of a long-forgotten emperor and a bygone civilization, of which only a few ruins remain, barely visible in the middle of a lifeless desert.

When civilizations were regional, the stakes were lower; one could pass away and the rest remain unaffected. Today, for the first time, the planet is dominated by a single global civilization—the modern civilization of science, technology, nations, and global consumers. When this one passes, as all eventually do, the consequences will be far greater. The U.S. government may believe that some banks are "too big to fail," but there is no power capable of bailing out a global civilization when it stumbles. Once again, for the 50th or 100th time in human history, the rhythmic cycle of civilizational change will begin again. The first sounds of that drum beat can be heard even now.

The next civilization—the next pattern of social organization—will be an ecological civilization if it is to be at all. The term "ecological" here does not express a utopian dream (see Question 3 below). At its minimum, it simply means whatever comes after our present civilization. Maybe there won't be another civilization; maybe humans will be living in trees, or maybe there won't be humans. But if there *is* to be any kind of stable society that endures when modern civilization has ended, it will have to be a sustainable one. That is, it will have to be enough in harmony with the environment that it avoids (or reverses) the kinds of devastating damage that modern civilization is now causing.

This is the great thought experiment of our day: to reflect on what kind of a civilization—what kind of overall world- and life-view—will arise when modernity has burned itself out (perhaps literally) and has given rise to its successor. What can the ecological sciences teach us about the kind of civilization we need to develop? What lessons can we learn by studying the painful mistakes of the modern period? Specifically, what have we learned about the consequences of global economic systems that make the rich richer and the poor poorer; the over-consumption and overuse of natural resources; the organization of societies for individual gain rather than for the good of the whole; and the reliance on energy sources that destroy the planet's ecosystems?

The theme of hope runs through this book. As we examine the natural rhythm of civilizations—how they rise and fall—a new form of hope emerges. For every civilization that has perished, something new and previously unknown emerged in its place. While we may lament the loss of the modern civilization and the comforts it carried, we can find hope in the new forms of life that will follow. Throughout the rest of the book we will attempt to show that ecological civilization is much more than a dream about the future—it is a guide to specific actions, both at the policy level and in the ways that individual people live.

Endnotes

1   Charles Taylor, "Modern Social Imaginaries," *Public Culture* 14.1 (Winter 2002): 92; italics added.
2   Taylor, "Modern Social Imaginaries," 92.
3   "Civilization," National Geographic Society, https://www.national-geographic.org/encyclopedia/civilization/.
4   For a list of civilizations from Mesopotamia to the Global Civilization, see "History of Civilization," HistoryWorld, http://www.historyworld.net/wrldhis/PlainTextHistories.sp?historyid=ab25.
5   See Sydella Blatch, "Great achievements in science and technology in ancient Africa" *ASBMB Today* 12.2 (February 2013): 32–33; http://www.asbmb.org/asbmbtoday/asbmbtoday_article.aspx?id=32437.
6   Drawn from "Mayan Scientific Achievements," History, last modified August 21, 2010, http://www.history.com/topics/mayan-scientific-achievements, and Anirudh, "10 Major Achievements of the Ancient Maya Civilization," published February 26, 2017, https://learnodo-newtonic.com/mayan-achievements.
7   University of Chicago 2017–2018 Catalog, under the heading "Civilization Studies," http://collegecatalog.uchicago.edu/thecollege/civilizationstudies/; italics added.
8   Nicholas St. Fleur, "Volcanoes Helped Violent Revolts Erupt in

Ancient Egypt," *The New York Times*, October 17, 2017, https://www.nytimes.com/2017/10/17/science/volcanoes-ancient-egypt-revolts.html.

9   Data from *Forbes* magazine, https://www.forbes.com/sites/christopherhelman/2016/06/28/how-much-electricity-does-it-take-to-run-the-internet/#4144148d1fff.

10 The Google Ngram viewer shows that the usage of "ecological" shot up before and after 1970, and then shot up again in the 1990s.

11 Holmes Rolston, "Science and Religion in the Face of the Environmental Crisis," in *The Oxford Handbook of Religion and Ecology*, ed. Roger S. Gottlieb (New York and Oxford: Oxford University Press, 2006), 387.

12 Richard Dawkins, *The Selfish Gene* (Oxford: Oxford University Press, 1976), xxi.

13 The first part of the quote is from Dawkins, *The Selfish Gene*, 11; the second occurs only in the 1989 edition of *The Selfish Gene*, 20.

14 See John B. Cobb, Jr., ed., *Back to Darwin: A Richer Account of Evolution* (Grands Rapids: Wm. B Eerdmans Publishing Co., 2008).

15 Robert Ulanowicz, formerly at the Chesapeake Biological Laboratory, speaks of a "third window," since ecosystem dynamics cannot be described in either Newtonian or Darwinian terms. See Robert Ulanowicz, *A Third Window: Natural Life Beyond Newton and Darwin* (Philadelphia: Templeton Foundation Press, 2009).

16 Lynn Margulis, ed., *Symbiosis as a Source of Evolutionary Innovation: Speciation and Morphogenesis* (Boston: MIT Press, 1991).

17 Martin A. Nowak, "Five rules for the evolution of cooperation," *Science* 2006 (Dec 8); 314(5805): 1560–63; doi: 10.1126/science.1133755.

QUESTION 2

WHAT ARE THE UNDERLYING CAUSES OF ECOLOGICAL CATASTROPHE?

HUMANS HAVE MADE THE PLANET SICK. And like any illness, simply treating the symptoms without understanding the root cause can be a fatal error. If we want to heal the planet, if we want to combat the threat of climate change, we need to address the underlying causes. So what are the deeper causes of our current climate crisis?

According to **NASA**, "Most climate scientists agree that the *main cause* of the current global warming trend is human expansion of the 'greenhouse effect'—warming that results when the atmosphere traps heat radiating from Earth toward space."[1] And while the greenhouse effect may explain changes to the climate, it leaves unanswered the question of the underlying cause of the greenhouse effect. Of course, most climate scientists would then turn to explanations of how the increase of certain gasses (like carbon dioxide, methane, etc.) block heat from escaping. Yet, even this explanation leaves unanswered the question of why there is an increase of CO_2 and CH_4 in the atmosphere. At this stage, we get a greater variety of explanations—from human population growth, to industrialized farming and the production of meat, to the increased use of combustion engines and general fossil fuel consumption . . . the list goes on. Yet behind each of these contributors to the climate crisis, there is another underlying cause.

We can follow this causal chain further and further back until we arrive at the bedrock of our civilizational structure—a set of big ideas, basic assumptions about the world that provide the foundational framework or paradigm on which our civilization was built.

Ecological civilization is first and foremost a form of civilization. It is a positive vision of a society built on the principles of sustainability and commitment to the common good, a world that works for all. But why would changes in the climate require changes of civilization? What does the environmental crisis have to do with economics, politics, or education?

The ecological civilization framework—the EcoCiv paradigm—is a living systems framework, in which our world is understood as an organic, dynamic, interconnected, complex system. This portrayal of reality may seem obvious, but it is actually not the historically dominant worldview. In order to understand why responding to the climate crisis requires a change of civilization, we need to better understand the big ideas (paradigms) that have contributed to the climate crisis.

Big Ideas That Have Changed Everything

As Victor Hugo said, "There is one thing stronger than all the armies in the world, and that is an idea whose time has come." Throughout history there have been a number of "big ideas" whose times came, and they changed everything.

- *Farming.* Modern-day farming can be traced back to more than 14,000 years ago. The development of agriculture allowed hunter-gatherer societies to form permanent dwellings—the foundations of civilization. With food no longer the sole motivation for existence, the way was paved for the development of government, trading, and individual specialization.

- *Electricity.* The ability to harness electricity is one of the pillars of contemporary civilization. Electricity has become a practical tool of modern life: from communications, transportation, and refrigeration, to commerce, computers, and medical equipment.

- *Germ Theory.* The germ theory of disease states that many diseases are caused by microorganisms that are invisible to the naked eye, but capable of invading living hosts. Proposed back in the 1500s and expanded 200 years later, germ theory provided the foundation for modern medicine, leading to the development of vaccinations and the practice of sterilization that has saved countless lives over the past few centuries.

- *Zero.* Although zero has been around since the time of Indic civilization, it didn't infiltrate Western thinking until the 12th century (thanks to Italian mathematician Leonardo Fibonacci). The concept of zero paved the way for the concept of decimals, allowing advancements from simple algebra, quantum physics, and rocket science to the binary code that is the basic language of all modern computers.

- *World Wide Web.* Speaking of computers: while the idea of connecting computers through networks dates back to the late 1960s, it was the creation of the world wide web some 30 years later that resulted in what is known today as the "internet," revolutionizing communications, commerce, and the dissemination of knowledge. Because the web is a relatively recent development, only time will tell the full impact of this revolution on human society.

Most of these ideas were initially met with great skepticism, even hostility. As Alfred North Whitehead states, "Almost all really new ideas have a certain aspect of foolishness when they are first produced."[2] Or, to take a sentiment attributed to Albert Einstein: he is quoted as saying, "If at first the idea is not absurd, then there is no hope for it."[3] The truth is, if even one of these five ideas hadn't prevailed, we'd be living in a much different world today! Each one has led to such monumental change that it can be called nothing less than a paradigm shift.

What is Ecological Civilization?

What's a Paradigm and Why Does It Shift?

A paradigm is a model, template, or archetype—a framework for understanding. Consider the image below[4] . . . what do you see?

Is it a duck? Is it a rabbit? Let's imagine you lived in a place that had lots of ducks, but no rabbits. Your "duck paradigm" would predispose you to see a duck. On the other hand, if you've never seen a duck before, but are familiar with rabbits, your framework for understanding the image would result in seeing a rabbit.

A paradigm shift is an important change that happens when the usual way of thinking or doing something is replaced by a new and different way. We used to only see rabbits. Now we can see ducks. Better yet, now we see the truth about the ambiguous figure—that it can be seen as either duck or rabbit. While this might seem like a strange, even unimportant example, our paradigms, our worldviews, and our fundamental assumptions about reality shape **EVERYTHING** we say, think, and do.

Wrong Paradigm—Wrong Direction

Consider a map. Imagine you're on a road trip from Los Angeles to the Grand Canyon, but only have a map of Beijing. How could you reach your destination using a map that didn't include your destination? When we know the world through the map alone, we are bound by its borders. If we, as a society, want to find our way to a desirable destination—such as a sustainable world—we need the right framework, the right paradigm, the right map.

Unfortunately, our modern civilization is built on the wrong paradigm. The "modern" map leads to environmental catastrophe. If we are to find our way to a more sustainable and just way of living on this planet, we will need a new map—a new paradigm.

But changing paradigms is not easy. It's not enough to "know" that we should be living differently. As stated in the movie *Inception*, "What is the most resilient parasite? Bacteria? A virus? An intestinal worm? *An idea*. Resilient . . . highly contagious. Once an idea has taken hold of the brain, it's almost impossible to eradicate." Therefore, if we are to change our paradigm and all that comes with it, we need to follow the advice of Yoda, who says "You must unlearn what you have learned." But before we can unlearn what we have learned, we need to examine what we have learned—acknowledging our most basic assumptions.

The Situation: Our Modern Paradigm

As environmental philosopher John B. Cobb, Jr. says,

> We must be honest. We live in a terrible time. We know that our actions are destroying the ability of the Earth to support us, but we seem incapable of changing direction. We plunge blindly ahead, either ignoring the reality of what is happening or hoping that some technological miracle will save us. It will not. The *modern world* has overshot the limits of what the Earth can bear, and our civilization will collapse.[5]

So how did we get here? What about the modern paradigm, the values and worldviews, that have led us to the brink of collapse? "Modernity," characterized by the "Enlightenment," is often traced back to René Descartes (1596–1650), the "father of modern philosophy." Descartes is best known for his statement "I think, therefore I am." However, it is what lies beneath that assertion that helped Descartes set the stage for the modern period—Cartesian dualism.

What is Cartesian dualism? It's the idea that there are only two things that make up reality: (1) matter in motion on the one hand, and (2) mind or the human soul on the other. For Descartes, everything falls into one of these two categories. Further, as a type of dualism, mind and matter are believed to be *independent* of one another. One has no bearing on the other.

Among other things, this dualism resulted in the elevation of humans above all other creatures. This results in an anthropocentric, or "human-centered" worldview. Humans are unique because they possess a soul or mind. For Descartes, this makes humans superior to other animals. Although this conclusion carried the alienation from nature to its extreme, it gave dignity to human beings. It supported the idea of human rights and even of a fundamental equality of all human beings, which is evidenced by the development of modern democracy.

When Charles Darwin later showed that human beings are a product of evolution (i.e., fully part of nature), this opened the door to a re-thinking of nature as having some of the properties Descartes attributed only to the human soul. But, by that time, the commitment of the sciences to methods associated with nature's purely objective existence was very strong. Instead of changing the approach to the rest of the natural world, most scientists chose to study humans in the way they had previously studied the objects of physics and chemistry. As a result, enlightenment dualism was replaced in late modernity by reductive physicalism, the idea that reality is *only* made up of the physical. No mind. No soul. Just matter.

This paved the way for the "objective" study of the world as a machine, without values or purposes—in short, for a mechanistic worldview. The quest for certainty, through objectivity, became the heart of educational systems and led to the creation of "disciplines." So dualism paved the way for the fragmentation of knowledge.

By breaking knowledge up into disciplines, fields of study like "economics" could emerge independent of the natural world. This resulted in a "fallacy of misplaced concreteness," a reification in which abstract economic theories such as unlimited growth are mistakenly treated as concrete facts. One consequence was the view that unlimited growth is a realistic possibility. This development is explained in detail by the ecological economist Herman Daly and the eco-theologian John Cobb in their book *For the Common Good: Redirecting the Economy Toward Community, the Environment, and a Sustainable Future*.

Question #2

The environmental crisis is (in part) the result of pursuing unlimited growth on a limited planet, which is (in part) the result of mistakenly treating the abstract principles of economics as concrete realities, which is (in part) the result of fragmented disciplinization, which is (in part) the result of the reductive physicalism that scientists adopted after Darwin. All of these mistakes can be traced back to a deep and often unconscious commitment to Descartes' mind-body dualism, which was the "big idea" that grounded the Enlightenment and the Modern paradigm.

It has now become clear that *Cartesian dualism, a central idea of the Modern paradigm, paved the way to our current environmental crisis.* Now, it's not as if neo-liberal economists recognize their practices as indebted to Descartes' philosophy. However, we are always partly constituted by the past. We have inherited a legacy from our ancestors . . . whether genetic, social, or other; whether recognized, ignored, or embraced.

The Climate Crisis Is a Civilizational Crisis: We Need a New Paradigm

If Descartes' philosophy serves as a paradigm that leads toward unsustainability and collapse, then we need an alternative to Descartes—a new paradigm that leads toward a sustainable and just world—an ecological civilization. If we have the wrong framework, the wrong paradigm, the wrong map, then no matter how hard we work, we won't reach an ecological civilization.

Among others things, ecological civilization involves the following paradigm shifts:

- from dualism (mind vs. body) and monism (only bodies exist) to holism (minds and bodies interwined);
- from mechanism to organism;
- from anthropocentrism to biophilia;
- from unlimited growth to sustainability.

So, what does ecological civilization affirm? It affirms that the changes required in response to global climate disruption are so extensive that they must be linked to another form of human civilization, one based on ecological principles. Ecological civilization will require a synthesis of economic, educational, political, agricultural, and the other societal reforms in order to achieve sustainability.

At an even deeper level, because environmental conditions are deeply tied to these reforms, achieving a sustainable civilization will involve a shift in awareness and values. It will involve a significant rethinking of current structures and practices in a more systemic way than has been true in the past.

In June 2015, a group of scholars, activists, and concerned citizens gathered in Claremont, California, for this very purpose—to explore alternative paradigms and practices toward an ecological civilization. This transdisciplinary conference consisted of some 1500 participants collaborating in roughly 82 working groups on various topics. While the event was open to people of all backgrounds and beliefs, the organizing body (the Center for Process Studies) was convinced that a family of thought known broadly as "process-relational philosophy" is perhaps the best alternative to the destructive modern paradigm.

Process philosophy of this variety originates with the work of mathematician and philosopher Alfred North Whitehead, who referred to his system as a "Philosophy of Organism." Leading process philosophers John B. Cobb, Jr. and David Ray Griffin also began using the phrases "Constructive Postmodernism" and "Ecological Civilization" as alternative ways of describing process-relational thinking—phrases that have gained much traction in China over the past few decades.

Process-relational philosophy can be summed up by three basic principles:[6]

1. No one can cross the same river twice: being is becoming. (process);
2. No person is an island: all things are interconnected. (relational);
3. Seeing heaven in a wildflower: all living beings have value.[7]

Whitehead needed a way to explain how something that is always changing can be fully actual. In response, he put forward the idea of

Question #2

"concrescence." Concrescence is simply the process of becoming "concrete." As John Cobb explains, "Concrete means fully actual, and that means a completed actual occasion [an entity]. The use of the term 'concrescence' places emphasis on the idea that even these momentary flashes of actuality that Whitehead calls actual occasions are processes."[8]

Now the process of becoming fully actual (becoming "real" or "concrete") also involves something Whitehead calls "prehension." Prehension emphasizes one's relation to the past as it contributes to the process of becoming what one is. We become what we become through our prehensions—always (at least partly) constituted by the past. Together, these two processes (prehension and concrescence) explain what it means for reality to be a process of interrelated becoming.

The third notion mentioned above, that all life has value, is a conclusion that arises from understanding the interconnected process of becoming. According to process philosophy, value is "inherent in actuality itself." The whole process of becoming is directed toward the grasping of value in others, incorporating it within oneself, and furthering value in creative ways. As Whitehead writes, "Our enjoyment of actuality is a realization of worth, good or bad. It is a value experience."[9]

Please don't misunderstand; we are not suggesting that, if one simply thinks the right things, this philosophy will magically minimize the greenhouse effect. After all, as Whitehead declares, "Ideas won't keep. Something must be done about them."[10] What we are saying, however, is that the root cause of the climate crisis should not be confused with its symptoms. The climate crisis is a civilizational crisis. As such, we need a new framework—a new paradigm—for a different kind of civilization: an ecological civilization.

The foundation of the modern paradigm was a philosophy of dualism that depicted reality as mind vs. matter, with each independent of the other; this dualism later reduced reality to physical matter-energy alone, void of purpose and value. Under the ecological civilization paradigm, dualism and monism are replaced with a holism that portrays reality not as a collection of objects, but as a community of subjects—an interconnected whole, within which we are constituted by our relations.

According to the standards of modern industrial agriculture, success is defined in terms of "productivity," which is measured by product divided by hours of human labor. Under the ecological civilization paradigm, where everything is interconnected, success in agriculture means regenerating the soil for sustainable farming.

In the modern paradigm, economic success is described in terms of growth. This has proven detrimental to our planet. Small steps, like embracing a "triple bottom line," are not enough so long as short-term gains on quarterly reports have priority. With an EcoCiv paradigm, we redefine economic success in terms of the overall well-being of people and the planet (the common good). This might include something like Ecological Economics, or even the Economics of Happiness.

According to the modern paradigm, the purpose of higher education is defined by specialized knowledge, career development, and preparation for high-paying jobs. Under the eco-civ paradigm, the role of education is to empower leaders to serve the global common good. This requires a system of education that does not attempt to be "value-free," but instead seeks to develop wisdom, nurture integrative knowledge, and promote the common good. In reality, the world cannot be neatly divided into disciplines—they all affect and bleed into one another. Nor should such neat but arbitrary divisions inform the way we learn about the world. We need an educational system that supports and values rural farmers as much as corporate businessmen.

Paradigm shifts in these and others areas would result in a radically new form of civilization. This new form of civilization—an ecological civilization that at its core makes a commitment to the common good and the well-being of the planet—is the only way we can address the underlying causes of the climate crisis. If we truly wish to heal the planet, addressing more than just the symptoms of climate change, we need an ecological civilization.

Endnotes

1   "Global Climate Change: Vital Signs of the Planet," NASA, https://climate.nasa.gov/causes/; emphasis added.

2   Alfred North Whitehead, *Science and the Modern World*, (New York:

The Free Press, [1925] 1967), 47.

3   The source is unknown, but (as someone quipped) the idea is so absurd that we can only hope he said it.

4   Duck/Rabbit drawing from Joseph Jastrow, *Fact and Fable in Psychology* (Riverside Press, 1900), 295.

5   John B. Cobb, Jr., "Ten Ideas for Saving the Planet," http://www.ctr4process.org/whitehead2015/ten-ideas/.

6   We acknowledge that, put in this brief form, this account oversimplifies the position, which brings inherent limitations and dangers.

7   Jay McDaniel, *What is Process Thought? Seven Answers to Seven Questions* (Claremont, CA: P&F Press, 2008).

8   John B. Cobb, Jr., *Whitehead Word Book: A Glossary with Alphabetical Index to Technical Terms in Process and Reality* (Claremont: P&F Press, 2008), 59.

9   Alfred North Whitehead, *Modes of Thought* (NY: The Free Press, [1938] 1968), 116.

10 Lucien Price, ed. *Dialogues of Alfred North Whitehead* (Boston, MA: Little, Brown, 1954), 254.

Question 3

IS ECOLOGICAL CIVILIZATION MERELY A UTOPIAN IDEA?

POSITIVE THINKING is a good thing, as is hope for a better future. Among the regions of the world, Europeans and North Americans stand out for their belief that, whatever the state of the world at a given time, they can make the world a better place. Many positive changes in the world can be traced back to this optimism.

Utopias, Dreams, and Realities

A *utopia* is a vision of a future that is far better than the present. Indeed, a utopian ideal often describes a *perfect* society. On the one hand, imagining an ideal society can move people's hearts and minds. Think of the famous song "Imagine" by John Lennon.

On the other hand, there is a dangerous side to utopian thinking that can be damaging, even disastrous. The early modern period in Europe is described as *melioristic*, which means the belief that things are getting "better and better." A classic example is the philosopher of G. W. Leibniz (1646–1716), who based his philosophy on the idea that we live in "the best of all possible worlds." Indeed, many of the Enlightenment thinkers in the 18th century held that reason reveals the path toward a better and better future, because when humans know the right thing to do, they will do it.

Part of the danger of utopias lies in what people don't see. For example, in the modern period things *did* get better and better for many Europeans . . . thanks to the colonizing of Africa, India, and South America, the massive reliance on slaves, the size of their armies and navies, and new technologies for killing more and more people more and more effectively. As Europeans began to colonize North America, they brought their belief in progress to the New World, where it morphed into a doctrine known as *Manifest Destiny*. It is the indisputable destiny of Europeans, the settlers proclaimed, to move across America "from sea to shining sea," killing the native peoples, taming the earth, and establishing what they believed was the superior European culture and religion. Progress was achieved through domination, which was inherently violent.

Famous Utopias

As we consider whether or not ecological civilization is merely a utopian ideal, it is helpful to bring to mind some of the most famous utopias:

Sir Thomas More. More invented the word utopia in 1516 when he published a book by that title. It was about an island off the coast of South America where a perfect society had been formed. Living conditions were ideal, and human suffering was minimized. The highest activities of philosophy and the arts were being pursued. Citizens were free, and justice reigned. Because the island didn't exist, he called it u-topia, which means "not a place." Because *u-topia* and *eu-topia* ("a good place") sound the same in English, a utopia came to mean a good place or a good state that lies ahead of us in the future.

The coming Messiah. The belief that the Messiah will come is an important strand of Jewish thought, though by no means the only one. Jews have been oppressed, imprisoned, and killed for millennia, unable to protect themselves from the anti-Semitism (and armies) of the surrounding peoples, most often Christians. Without a homeland and surrounded by injustice, Jews have found that the promises of the Messiah who is to come have been left unfulfilled. But someday, the Orthodox tradition holds, God will send the promised Messiah, who will establish justice and peace, punish evil-doers, and re-establish Israel, the Temple in Jerusalem, and the promises made to Abraham.

The Christian Millennium. Likewise, one tradition in Christian theology (*millenarianism*) held that, after the Second Coming of Christ, there would be a 1000-year reign of Christ on earth. There will be no war or suffering in this golden age; justice will reign, and the earth will be filled with peace and prosperity.

Marx. In the *German Ideology* Marx demonstrates the harm caused when the market (as we call it today) forces people to accept and retain a particular role in the workplace, lest we "lose [our] means of livelihood." In a famous passage he contrasts the way that the capitalists (corporations) control workers with the way we could live without capitalism:

> in communist society, where nobody has one exclusive sphere of activity but each can become accomplished in any branch he wishes, society regulates the general production and thus makes it possible for me to do one thing today and another tomorrow, to hunt in the morning, fish in the afternoon, rear cattle in the evening, criticize after dinner, just as I have a mind, without ever becoming hunter, fisherman, herdsman or critic.[1]

In each of these examples, a utopia represents a way of life that (it is claimed) maximizes peace and prosperity while minimizing suffering and injustice.

The Dangerous Side of Utopian Thinking

The flip side of utopian thinking expresses itself in several ways. First, because people see the final outcome as perfect, they are inclined to use all means possible to achieve the goal—a classic case of "the end justifies the means." Second, cultural blindness brings with it the threat that we will impose our own values, conceiving utopia only in terms of our own culture and values. Third, utopias are often unrealistic: they don't encourage people to do the concrete things they *would* do to make the world better if they were not so preoccupied with attaining the perfect society. Fourth, utopian thinking may be defeatist. If everything is determined to bring about the world of the Messiah or the Second Coming, why should I work for it? Perhaps the most famous example is Calvinism, a branch of Protestantism that holds that God is completely in charge of history, so that nothing humans do can change the outcome.

Finally, one's utopia may be misguided. Nations such as East Germany interpreted Karl Marx as requiring that all large businesses be owned by the state. But when industries were put under state control, they became inefficient; corruption replaced hard work. The collapse of the East German economy eventually led to the reuniting of the two Germanys as a single capitalist state. When people work toward a mistaken goal, the consequences are often deadly.

We have to ask: is ecological civilization a utopia in this sense?

Utopia or Dystopia?

Today we are less apt to dream of utopias than we are to worry about *dystopias*. A dystopia is "an imagined place or state in which everything is unpleasant or bad, typically a totalitarian or environmentally degraded one" (*Oxford Living Dictionaries*). Instead of believing that "things are getting better and better," many today hold the view that—as the old saying goes—"things are going to hell in a handbasket."

Dystopian books and films have become extremely popular; they express the mood of our age. From television shows like *The 100* to movies like *Mad Max* and *The Day After Tomorrow*, dystopian visions of the future dominate Hollywood. The unbelievable popularity of zombie films—we think in particular of *The Walking Dead*—represents without a doubt the clearest and most compelling example. Recall Cormac McCarthy's immensely popular book, *The Road* (2006). A father and a boy are walking through a dead world, along an endless road. A nuclear war, presumably, has killed all crops and trees and almost all people. Those who remain can only hunt for hidden stores of food in deserted houses or kill each other and take what they have. Carrying all their remaining possessions in a grocery cart, the boy and his father struggle their way through the endless grey over mountains and across vast valleys. Nothing lives; nothing brings hope. Or think of the end-of-the-world novel, Karen Thompson Walker's *The Age of Miracles* (2013). The planet is slowing down; each day becomes longer. Bit by bit, life as we knew it collapses: the birds die, groups fight each other; people start to die of mysterious illnesses. There is no hope of returning to the old, normal existence; humanity is helpless as it faces the coming disaster. The parallels with global climate disruption are only too clear.

Religions can also create dystopias—for the planet, if not for the ones who will be saved. For example, when Ronald Reagan became president, he hired neo-conservatives to write position papers that would help determine U.S. federal policy. One paper, later leaked to the public, argued that the president should not invest in protecting the forests. The reason was that Jesus would return soon, so long-term forest preservation did not need to be a priority.

One might say that the true mark of dystopias is not simply the assertion that "times are tough," but the framework of "us vs them." Dystopias tend to prioritize survival, which is typically framed as individualistic. If there are not enough food and supplies to go around, sharing isn't an option. This means that theft, violence, and distrust of others are central themes in dystopian visions. Unlike dystopian thinking, ecological civilization thinking suggests that even when times are tough and resources are scarce, our best bet for survival is cooperation rather than competition.

Utopian thinking can offer a rosy picture of the future that is not based on actual data and trends. Because utopias are unrealistic, such dreams of the future fail to spawn useful action. At the other end of the spectrum, many responses to the environmental crisis are very specific and short term. They guide decision making in the near term, but they fail to provide hope or guidance for the longer-term future.

Fortunately, there is space that is not so close that we lose perspective, nor so far into the future that we lose knowledge. Ecological civilization, in the sense we are describing it here, functions as that space where both perspective and knowledge are possible. It presupposes that the current crisis is so great that it will require civilizational change, and that whatever social organization follows our consumption-based society will have to be based on ecological principles. But instead of affirming a carefree utopian world, it serves as a realistic framework for selecting among current responses to the crisis. And, perhaps most fundamentally, it offers hope that humanity can move beyond the flawed principles that underlie and justify modernity.

Social Imaginaries

Social imaginaries express "the set of values, institutions, laws, and symbols common to a particular social group and the corresponding society through which people imagine their social whole" (Wikipedia). The idea has become extremely influential in the years since Charles Taylor published his important book, *Modern Social Imaginaries*.[2] Taylor uses the term to draw attention to the particular (and often unconscious) ways that members of a large culture or civilization construe their collective social life together. He writes,

> The social imaginary is not a set of ideas; rather it is what enables, through making sense of, the practices of a society ... Central to Western modernity is a new conception of the moral order of society. At first this moral order was just an idea in the minds of some influential thinkers, but later it came to shape the social imaginary of large strata, and then eventually whole societies. It has now become so self-evident to us, we have trouble seeing it as one possible conception among others.[3]

Taylor, one of the great intellectual historians of the modern period, brings modernity's core values to the surface. As we recognize the global disaster that these values have created, we begin to imagine different options—sustainable options—for human life on this planet that will lead in radically different directions. These new goals are "actionable"; they offer a different kind of guide for policy formation and political action.

Human social imagination is a very powerful tool. Sometimes it produces vague utopian dreams for the future, and sometimes it expresses the hopelessness of dystopian views of the world. In the best cases, however, social imagination can be a key to unlocking visions of the future that are worth working for.

Utopias can also be used in a specific and constructive way; for example, as sketches of a future society that has learned to live in sustainable ways with the Earth's ecosystems. Ecological civilization falls in this category. It begins with a harsh realism: the goal of long-term sustainability cannot be achieved within the context of global civilization as it currently exists. Nation states failed to take sufficient action as the time ran out for avoiding the most disastrous effects of climate change, business and educational institutions have made only minor changes, and consumers are increasing rather than decreasing their patterns of consumption. The pursuit of an ecological civilization then requires a second and equally demanding form of realism. It mandates concrete reflection on specific economic and social systems: forms of housing and transportation, of production and consumption, of government and business, even of religion and art, that will be genuinely sustainable over the long term.

In short, unlike most utopian thought, ecological civilization is preoccupied as much with the means—the radical steps that humanity must take—as it is with the overarching goal. The goal of a world that works for all humans, living beings, and the planet as a whole is appropriate; but one should not therefore conclude that ecological civilization is all or nothing. As the French philosopher Voltaire once wrote, "The best is the enemy of the good." Ecological civilization is a vision for a *better* world, even though it will not be a perfect world. This distinction allows ecological civilization to resist the pitfalls of utopian ideals while retaining one of the core motivations of utopian thinking—hope.

Endnotes

1   Karl Marx, "Private Property and Communism," in *The German Ideology (1845): "Part 1A: Idealism and Materialism,"* para. 2; available online at the Marxists Internet Archive, https://www.marxists.org/archive/marx/works/1845/german-ideology/ch01a.htm.

2   Charles Taylor, *Modern Social Imaginaries* (Durham, NC: Duke

University Press, 2004). Taylor was influenced by an important earlier work, Benedict Anderson's *Imagined Communities* (London: Verso [1983] 2016). See also Appadurai's *Modernity at Large* (Minneapolis, MN: Univeristy of Minnesota Press, 1996), and Warner's *Publics and Counterpublics* (Brooklyn, NY: Zone Books, 2005.

3   Charles Taylor, "Modern Social Imaginaries," *Public Culture* 14.1 (2002): 91, 92.

Question 4

WHAT ARE THE FOUNDATIONAL INSIGHTS OF THE ECOLOGICAL CIVILIZATION MOVEMENT?

THE PHRASE "seizing an alternative" well expresses what draws many of us to focus on the longer-term goal of building an ecological civilization. Living as humanity has lived for the last 200 years clearly is no longer an option. Baby steps toward change—buying a Prius every three years instead of a SUV, or shopping for "green" products at the local supermarket—are too little too late. The global crisis can be addressed only by changing the system altogether. But what do the alternatives look like?

Let's explore the results from a particularly significant conference of scholars and activists. Some 1500 delegates gathered at Pomona College in Claremont, California, in June 2015 to explore the theory and practice of ecological civilization. Keynote talks featured leading environmentalists from three of the world's largest countries: Vandana Shiva of India, recipient of the Right Livelihood Award; Sheri Liao of China, founder of the Global Village of Beijing; and American environmentalist Bill McKibben, the founder of 350.org. Some 82 working groups, divided into 12 major sections. Each met for a total of 12 hours over the course of 3 days, and a number continued their work for months or years after the event.[1] Most of the leading scholars and activists in the field were present, not only from the West, but also from China and a number of developing countries.

Reviewing the causes of climate change is important, since it sets the context. Outlining the specific lifestyle changes that individuals and communities should make also helps. But neither is enough. The "Seizing an Alternative" movement has the far larger goal of studying every aspect of ecological civilization: its historical background, historical parallels, philosophical foundations, political and economic systems, international collaborations, and the concrete steps that must be taken to reach the goal.[2] What will it look like for humanity to come out on the other side of the ecological crisis? How will we have to live with the planet and with each other if human culture is to survive? If one doesn't understand the long-term goal, it's impossible to take concrete steps to get there. Is there an alternative to the late modern global market, and if so, what must be its features?

Five years in the making, the conference was the brainchild of Prof. John B. Cobb, Jr., whose overview of ecological civilization opened this book. For his motivation, Cobb turned to the organic philosophy and worldview of Alfred North Whitehead as a guiding framework. In his unpublished analysis of the results of the conference, Cobb writes, "I hoped to communicate to people in many fields of thought and activity that Whitehead's philosophy shows how they are partners in working for a single goal, which could be called ecological civilization."

The Seizing an Alternative movement shows what has to happen when a common quest toward an ecological civilization becomes the primary focus. It gives rise to a common research project and a palpable sense of commonality among a diverse set of scholars and activists who connect around a hopeful vision in the face of an overwhelmingly difficult task. This vision is ambitious; in Cobb's words from his unpublished summary, it is

> that many people will feel assured that their sense of sharing in many different ways in a common purpose is fully grounded intellectually and spiritually and that as a result cooperation and mutual support can replace fragmentation among those who are seeking to respond to the global crisis.

Question #4

The Seizing an Alternative conference was, we believe, the first time that the theory and practice of ecological civilization were worked out in systematic and rigorous detail, and no other events in the intervening years have been held at the same scale. In a new book, *Putting Philosophy to Work: Toward an Ecological Civilization* (2018), major keynotes from the conference have been published, providing an overview of ecological civilization as a global paradigm.[3] In this chapter, we draw on the unpublished conference proceedings as an introduction to the field of ecological civilization studies, exploring the foundational insights of the ecological civilization movement. We concentrate here on the broader paradigm in the attempt to better convey the common underlying message; more specific detailed will be covered in subsequent chapters.[4]

The Central Ideas

The nature of the Seizing an Alternative project can be boiled down to three central ideas. First, there must be a disciplined process for mapping an alternative future. The point of departure is the current environmental crisis, which requires at the very least a major commitment to work to create a sustainable society. Although sustainability is a required step toward an ecological civilization, it's not enough. The concept of sustainability is frequently used by individuals, organizations, and governments as a cover for business as usual, but with a nice PR twist. It is often used to give the appearance of change, but without genuine transformation.

In fact, the co-opting of sustainability by those who are profiting from the current system points to a far deeper issue about the way that people think and behave. Green band-aids are not enough. This leads to the second central idea: the birth of new civilizational structures is associated with the emergence of a new story or paradigm that expresses a different understanding of our place on this planet. Nothing less than a change of world- and life-view will suffice. It takes changed thinking to change the world. Radical changes in society require equally radical changes in the way we think, which means a change in worldview and philosophy (see Question 2 above).

Decisions about how humanity structures life and society today will have a greater effect on the future of the planet than any other factor. That's why scientists are now calling the current era in the planet's history the *Anthropocene*.[5] Given the level of human control over the planet's history, we cannot simply turn back to biocentrism, the view that humans are just one of many influences on the global biosphere. The only long-term alternative to this destructive human civilization is a civilization that is truly ecological—sustainable in the strongest sense of the word.

The starting point, then, is to build lifestyles, societies, and economies based on *the ecological paradigm*—the recognition that all things, both human and natural, are interconnected. Only after the realization that *all things exist in community*, rather than in static mechanistic relations, can new forms of society grow. In this way one begins to see in detail exactly how the ecological sciences help to ground an ecological worldview.

In a later chapter, we turn to the third basic idea: Changes in world- and life-view must lead to changes in one's life and practice. After the paradigm shift in thinking come the concrete details. Scientists, scholars, and activists are beginning to formulate what human society will actually look like as truly ecological practices take us beyond the merely sustainable (or the blatantly unsustainable!) to genuine transformation. Here "ecological civilization" moves beyond abstractions and becomes an action word. Proposed changes can now be evaluated in terms of their implications for rethinking and restructuring every sector of society: economy, culture, religion, education, energy and transportation policy, the revivification of rural life, the ecological transformation of urban centers, and much more.

Three related topics thus demand our attention: (1) our present context, (2) the needed changes in worldview, and (3) the first sketches of how to begin building an ecological civilization.

THE CONTEXT: BREAKTHROUGHS IN THE
STUDY OF ECOLOGICAL CIVILIZATION

The framework for thinking about ecological civilization includes global climate disruption, the resulting threat of social and economic collapse, and the changes that will have to follow in the wake of these events. Some people suppose that an ecological civilization is merely another word for sustainability. A sustainable society only becomes a serious possibility, however, when people fully understand the urgency of the threat.

Put differently, the broader context requires a three-fold understanding: of environments, of the principles of sustainability, and of all the dimensions of the current crisis. Dr. Martin Luther King, Jr. often ended his talks with the phrase, "The arc of the moral universe is long, but it bends towards justice." By contrast, Bill McKibben noted, the arc of the Earth's biosphere is short, and it bends toward heat. Time is not on our side: "If we don't get this done soon then we don't get it done."

The Ecological Civilization Debate in China

Nowhere has the goal of achieving ecological civilization been so explicitly stated as in China. At the 17th Congress of the Communist Party of China in 2012, the government explicitly accepted the goal of building an "ecological civilization." The goal, they said, is to form "an energy and resource efficient, environmentally friendly structure of industries, pattern of growth, and mode of consumption."[6] According to philosophy professor Yijie Tang at Peking University, "Process philosophy or 'constructive postmodernism' criticizes binary thinking and views nature and humans as an interrelated bio-community. This idea has important implications for the solution of the ecological crisis facing us today."[7]

Many Chinese scholars have presented the successes and challenges that China has faced over the last half-dozen years. One often reads accounts of China's successes, but also, increasingly, its temptation to follow the patterns of Western modernity. Kang Ouyang, a leading Marxist scholar in China and vice president of Huazhong University of Science and Technology, has explored the positive and negative sides of modernization. For him, the negative elements of late modern capitalism include treating economic growth as the ultimate goal;

neglecting ecology, the value of tradition, and aesthetic wisdom as complements to scientific knowledge; rejecting the positive roles that religion can play in human life; and overemphasizing individuality at the expense of community.

When it comes to human relations with the natural world, we have to recognize the extent to which globalized modern civilization, with its pattern of exploitation and profit, has become the driver of the crisis. Corporations and powerful nations assert themselves as the dominant players, destroying populations and ecosystems as byproducts of the profit-based (capitalist) system.

Although the author Wang Jin treats environmental issues primarily as legal issues, they are also political, social, and philosophical in nature.[8] Non-legal factors are also important, including the role of capital, interest groups, short-term planning, and the model of "development" espoused by the World Trade Organization (WTO) and the International Monetary Fund (IMF). The current environmental crisis has become the first indisputable alarm that is raising global awareness. The situation of natural systems—deforestation, loss of arable land, water shortage, toxic air, and the destruction of entire ecosystems—shows the connection of the environment with sustainable policies and practices. In short: the environment depends upon sustainability, and sustainability depends on the ultimate goal of reaching ecological civilization.

How Is Ecological Civilization Different from Sustainability?

"Ecological" does not mean merely "sustainable." Sustaining the current civilization is not enough to solve the crisis that our planet now faces, because keeping current systems going can be accomplished without profound changes. Most people who are currently comfortable—the wealthy and those who possess power in politics and business—have a personal interest in maintaining the status quo. They happily express their support for sustainable practices, as long as these do not demand that they change the basic way they think and behave.

Fortunately, the number of people who recognize that the status quo is no longer viable is growing. Given the exploding climate crisis, a civilization that can thrive over the long term will be profoundly different from any civilization that exists today or existed in the recent past. It will be "ecological" in the sense that human activities will be related to the patterns of the nonhuman world in ways that enable the whole planet to flourish.

Some say that a *civilization* cannot be *ecological*. But judgments about the alleged incompatibility of these two terms are actually statements about how society is *currently* structured, not about how it *could be* structured. "Civilization" in the modern sense is an expression of how we do things in the present or near future. Modern civilization serves the interests of those who currently profit from economic globalization; it is therefore unable to express the inherent contradiction between modern globalization and the requirements for planetary survival. The evidence shows that far more radical changes are required. Humanity must fundamentally alter the way it thinks and behaves, seeking to learn from nature rather than to master it. The time is short.

So is ecological civilization an oxymoron? The accelerating destruction of the systems that sustain us is emblematic of deeper problems, and the severity of these problems shows the need for urgent responses in a wide variety of areas. Humanity's mistakes are now written across the surface of the earth, within its oceans, and in its skies. Given the scope of the global effects, the notion of *civilizational crisis* best expresses both the current problem and the kinds of responses necessary.

Concretely, the environmental crisis is the product of societies structured by a logic of growth, human-centric thinking, and the prioritization of atomized individuals over communities. Despite the urgency of the situation, the environment is not at the top of the list for most governments, and many are taking damaging steps backwards. The problems manifested in the worsening environmental crisis express—sadly—the real priorities of modern civilization as a whole.

Some people's response to the situation is to assume that the only way to live ecologically is to undo civilization altogether to return to something like Rousseau's pure state of nature. It is indeed possible that all human civilization will self-destruct and that the few survivors will reorganize into competing hunter-gatherer tribes similar to the ways our ancient ancestors lived. But such hopeless scenarios are as defeatist as they are unjustified. Effective steps toward genuine transformation come from people who have not given up on the prospects for human culture. That there has never been a truly ecological civilization does not mean that there cannot be one in the future. So there is no reason to suppose that the very goal is an oxymoron. We can learn from natural and human history what an ecological civilization will need to be, and how to create one.

Ecological Civilization Requires a Radical Change of Philosophy and Worldview

In contrast to those who resist including "ecological" and "civilization" in the same phrase, we favor keeping both. Together they call attention to the importance of finding ways to organize human civilization so that it can endure. By contrast, sustainability has become a largely negative term: it now connotes primarily avoiding the collapse of existing systems. But hope depends on also having attractive goals and concepts that point to healthy, symbiotic relations with other creatures and with one another.

The leaders of the Seizing an Alternative movement are working to specify the concrete differences between the emerging ecological worldview and the dominant practices and ideas that shape our contemporary world across all its sectors: education, science, technology, economics, etc. The juxtapositions are revealing: organic *versus* mechanistic, communal and ecological *versus* individualist, relational *versus* substantial, partially self-determining *versus* deterministic. The goal, as Thomas Berry beautifully describes it, is to live in a community of subjects rather than a collection of objects.

In short, ecological civilization requires not only a fundamental transformation of modes of production and a different model of development, but also a fundamental transformation of worldview, values, and lifestyle. It will take a *post*-modern worldview, accompanied by the appropriate actions and structures, in order to bring about this transformation.

Question #4

TOWARD A NEW PARADIGM: CHANGES IN WORLDVIEW

The Task

Turning back to the earth has been particularly difficult for members of modern civilization. Too many believe that our philosophies, religions, and sciences will lead us ever onward and upward. Modern thought has imagined humans as minds transcendent of bodies, or as subjects mastering a world of objects, or as purely physical organisms responding to a purely physical world. These sorts of abstractions have had very concrete and damaging effects.

Often religious traditions have contributed to these myths. By turning attention up and away, in the delusion that we can transcend the earth, modern religions have helped to produce a catastrophic down-turning. It may be that this irony will give rise to a reverse irony: a catastrophic turning back to the earth in a way that does not wipe out living beings, but transforms humanity in the process.

What we need is not fewer philosophers, but a way to free philosophy from philosophy. For one, we need to move beyond the boundaries of philosophy understood in the limited sense of an academic discipline. That means freeing philosophy from the controlling hands of academic philosophers and returning it to the people. After all, philosophy is not just for philosophers. When philosophy is freed from philosophy, it becomes a tool for everyday life—a tool that can assist average citizens in challenging the core assumptions that undergird our present society. Such a philosophy is free to explore alternative types of worldviews and civilizations that are well outside the purview of modern presuppositions.

In short, moving toward an ecological civilization involves changes to our core philosophies about ourselves and our world. Modernity has achieved huge advances in science and technology, but it has not prized wisdom. Today its very advances are threatening to destroy us. Given humanity's new technological prowess, wisdom has become more important than ever. We need to renew philosophy as the love of wisdom (*philos* + *sophia*), recognizing that wisdom has been too little sought by the moderns who call themselves philosophers. We now have to look elsewhere for guidance.

The History

The word *science* was derived from the Latin *scientia*, knowledge, linked with *scire*, to know. Traditionally, acquiring knowledge did not mean reducing an area of study to more fundamental laws; indeed, good knowledge was systematic and comprehensive.

The success of the early empirical sciences, which were mechanistic, gave birth to a new paradigm of knowledge as reductionist and mechanistic. One thinks of Francis Bacon's posthumous book *The Masculine Birth of Time* (*Masculus Partus Temporum*) of 1603. The subtitle of this book on science and religion reads, "the greatest region of man's dominion over the universe." By implication, time was feminine before man's dominion because time was based on partnerships. As Bacon wrote, "[Knowledge's] objective is to bind, and place at your command, nature with her offspring." It is no accident that this new way of defining knowledge was occurring at the same time that European civilization was building empires around the world. The new model of knowledge needed to justify the power and dominion that the Europeans were exercising. For example, in contrast to the notion of knowledge as partnership, the colonizers needed to work with the (obviously false) assumption that the lands they were usurping were empty. Non-white peoples were subhuman ("uncivilized"), and natural systems were understood mechanistically by the new scientists.

The famous 17th century chemist Robert Boyle, who was for a time the governor of New England, reflected the same ideology of reducing living systems to mechanistic parts. For him, the idea that we should have veneration for nature is an impediment to establishing "the empire of man over the inferior creatures of God." [9] Unfortunately, this mechanistic picture of life fails to see that it is these "lesser" creatures, from bacteria to the higher primates, who are the heart of the ecosystems on which our lives depend. Ecologists now recognize how much life is defined by the interdependent diversity of the species at all levels, from the invisible micro-organisms in the soil to the pollinators that plants need in order to reproduce: ants, flies, bees, beetles, birds, and butterflies. Similarly, the success of the social order does not just depend on the male who "forces nature to do his will"; the social order is also a web in which the role of so-called low-level workers, indigenous peoples, women, and many other invisible members of the social system are equally important.

Gaining historical perspective is crucial if we are to overcome the reductionist idea of knowledge that grew out of early modern scientific thinking. In the end, "ism's" matter. It turns out that the way a civilization is structured is intimately related to the way its members conceptualize the world. Each set of assumptions in modern intellectual history has played its role in the environmental catastrophe we now face: the scientific revolution, the Enlightenment, modernist assumptions, global capitalism, and consumerism.

The mechanistic construction of the world, which found itself unable to encompass the whole of human experience, was then imposed politically and economically. Governments sided with the power of science and its technological offspring, and global economies sought to reduce humans to consumers. The history of the 20th century shows increasing fissures in the mechanistic model; as they widened, large militaries and their "police actions" became necessary in order to prop up the largest economies. In this sense, we live in a global order built on "power/knowledge" (Michel Foucault)—a regime of knowledge based on violence.

The only effective antidote is to challenge the modern paradigm of a world of mere "matter in motion," which can be known only by reducing all that is to fundamental physical laws. One can then return to philosophy as the love of wisdom. Wisdom exists outside the academy as much as within—and perhaps far more. To move it back to the center allows for the liberation of philosophy, the returning of thought to the people.

Paradigm Shift: The Example of Whitehead as
a Philosopher of Ecological Civilization

With his "philosophy of organism," Whitehead advanced a complex ecological philosophy that emphasized becoming, interrelatedness, and inherent value in all things. Taken as a guide to the ecological movement, his contributions can help to overcome some of the major obstacles that stand in the way of the transformation toward ecological civilization. As a framework of thought, they supply some of the conceptual links between vision, worldview, and action.

Whitehead was one of the early philosophers to offer a profound understanding of the unity and connectedness of all things, which he called a "philosophy of organism." Today "ecological" has come to express a similar understanding of the world. As people recognize the ways in which contemporary culture externalizes nature and exploits it for human benefit, it becomes increasingly clear that until we fundamentally change our way of perceiving ourselves and the world, humanity will continue to pursue suicidal policies.

A Whiteheadian philosophy of ecological civilization offers an alternative. An ecological civilization is one in which human beings are fully aware of themselves as part of the natural world, and aware of their close kinship to the other creatures who constitute that world. To regain this knowledge requires a conceptual framework that fully overcomes the dualisms of modern thought: mind *versus* matter, human *versus* nature, and value *versus* value-free.

It's impossible to make this transition if one still works with a notion of God as Supreme Being and the Creator of souls in an otherwise mechanistic world. Whitehead thus challenged the dualism of God and world. As he imagined it, God's aim in all things is the realization of value for the sake of the creatures and as an ongoing contribution to the divine life. This Whiteheadian vision encourages ecological thinking in ways that belief in an unchanging God could not do.

Promoting the intrinsic value of all life, including non-human life, lies at the heart of an ecological civilization. This conception reverses the exclusion of values from the natural world that has characterized modern Western thought, economics, and politics. Defining human and nonhuman life in terms of a valueless nature opens the door to treat them as means to an end rather than as ends in themselves.

Whitehead's philosophy challenges such value-free approaches. Consider the field of education. Research universities have vastly increased the amount of information available to humanity. By contrast, they have provided insufficient guidance on how this information should be used. They offer too little criticism of the assumptions of the modern world that have led to the excesses that now threaten the planet. Value-free universities that are based on silos of specialized disciplines, and the financial models that sustain them, do not motivate faculty and students to engage in research about the transformations of society and economics that are needed for attaining genuine long-term sustainability. By contrast, the Whiteheadian vision calls for an education oriented toward wisdom.

Whereas recognizing the plurality of aesthetic values in the arts need not threaten social order, the relativity (or absence!) of values does. Shifting in the direction of an organic philosophy will enable people in many cultures to appropriate values that lead toward an ecological civilization without focusing on specific sets of do's and don'ts.

For Whitehead, "the simple craving to enjoy freely the vividness of life"[10] is an aesthetic aim and involves an aesthetic process. The production of art is one expression of this drive. At a deeper level, Whitehead was concerned about the production of life itself. To hope for beauty is to hope for a world that explodes into an unimaginable richness of forms, a world that is both biologically diverse and culturally diverse, a world of endless creativity. This organic worldview offers a path of return, recognizing the fundamental value of living things and affirming their emergent complexity, diversity, and richness. It invites humanity to prioritize wisdom over specialization and fact-gathering. The affirmation of these goals spawns the hope that humanity can again support the coordination of life-form to life-form. Starting with the intertwined nature of each individual life, its creativity and enjoyment, one can extend these values to broader and broader structures of relations.

Mechanism versus process and interconnection. The problems humanity faces are not merely technical problems. The thrust toward economic globalization, for example, is rooted in a mechanistic, domineering, ecologically unsustainable worldview. The survival of the planet turns on leaving this worldview behind.

Consider the contrast. The traditional notion of fixed substances spawns either/or thinking: if one person or institution increases in power, the others must lose power. Process thinking argues instead that no individual can have a long-term influence except through cooperation. Increasing the power of cooperating agents increases the power of the system that sustains them all. The most important form of power is therefore the one that empowers others. Only when this dynamic is pursued are systems genuinely sustainable. If our great-grandchildren are to have any reasonable chance of inhabiting a welcoming world, they and their society will have to affirm their organic interconnectedness with other living beings and have left the mechanistic worldview behind.

The famous clock in Strasbourg served as the model of a mechanistic worldview in which past and future could be calculated. Today, the *Pando populus* grove in Utah, where thousands of trees grow up from a single root system, has become the model of the alternative, organic worldview.[11] If people can learn to understand themselves and their world using the Pando model, they will naturally begin to develop different goals for relating to each other and to the Earth, affirming different priorities about what is valuable and important. A civilization becomes "ecological" when it begins not with objects in isolation, but with interacting systems. When these interactions are healthy, the whole community of life flourishes.

To "seize an alternative" means to replace the old worldview with new ecological imperatives. For example, human beings need to respect the natural ecology and adjust their actions to it; and human relations must grow out of mutually supportive interconnections with one another. Modern thought may have sought to collapse wholes down to their parts, but a civilization is only sustainable when ecological relations are returned to the center.

Summary: Nature and Ecology

Before the rise of city-based cultures, a much smaller population lived in a relatively sustainable way with other animals and the natural world. We have much to learn from this phase of human history. Today's disastrous treatment of the natural environment expresses a profound alienation from it, one that is both intellectual and spiritual.

The new systems sciences offer compelling data on the consequences of human actions to the ecological systems on which we depend. These sciences unearth the real relations in the natural world, which exists as systems of systems of systems. Organisms continually evolve; complex natural systems self-organize; new kinds of agency emerge; actors and ecosystems are fundamentally interdependent.

Decades of empirical studies reveal ecosystems to be tapestries in which all the threads are interwoven. Indeed, our fate hangs by a thread; to remove ourselves from the systems of life, and to destroy them, is to destroy ourselves as well. To "seize this alternative" is to call into question the reigning paradigm within which science today is being interpreted: anthropocentrism, binarism, dualism. It is science itself, and not idealistic thinking, that challenges these old paradigms.

An ecological civilization will not need to give up every action that modifies nature, but it will need to learn from nature and from its ability to create ecosystems that increase in complexity and richness over time. Of all the limitations of the modern worldview with which the natural sciences have so strongly allied themselves, the one with the most immediate relevance to our survival is excluding from scientific explanation the distinctive characteristics of life. Without understanding the dynamics of living systems, how can one protect them?

Ecological civilization studies, as represented by the Seizing an Alternative movement, is a genuinely new field of study that involves bringing big ideas (philosophies and worldviews) to work in the world through concrete actions and policies. Because it is about civilizational change, the movement touches on all aspects of society. The intersections of theory and practice, global and local, environmental and social, scholarship and activism are among its central features.

Ecological civilization is more than sustainability, but it must be sustainable. It's more than environmentalism, but it must involve living in harmony with nature. It's more than a philosophy, but must involve a change in worldview. As a paradigm for living in harmony with one another and the planet, it emphasizes the inevitability of a comprehensive transformation of human civilization, top to bottom. That isn't to suggest that modern civilization needs a fire sale, where everything must go; there are aspects of contemporary society that we can and should retain. Nevertheless, the ecological civilization movement is about moving toward a new reality—seizing an alternative. It's a vision of hope for a better future.

While China has been among the leading voices in ecological civilization, the movement has now expanded globally. Over the last few years a variety of other movements have emerged—movements that promote very similar visions while using distinct but compatible language and values. It is to these parallel movements that we now turn.

Endnotes

1   A number of volumes with conclusions from the Seizing an Alternative movement have been published as part of the "Toward Ecological Civilization" series of Process Century Press.

2   See the excellent report on the conference by Herman Greene at http://ecociv.org/the-great-conference-at-claremont-re-imagining-civilization-as-ecological/. Greene covers the events and content in more detail than any account so far.

3   See John B. Cobb, Jr. and Wm. Andrew Schwartz, eds., *Putting Philosophy to Work: Toward an Ecological Civilization* (Anoka, MN: Process Century Press, 2018).

4   The material in this chapter is based on transcripts and recordings from many of the 80 working groups and from hundreds of sessions. Credit for the texts and ideas that follow goes solely to the 1,000 or so working participants. Only in this way can we present in just a few pages what can be achieved when one begins to work seriously to spell out the nature of ecological civilization, and the first steps that lead in that direction.

5   See the bestselling book *The Sixth Extinction: An Unnatural History* (New York: Picador, 2015) by Elizabeth Kolbert; also Jeremy Davies' *The Birth of the Anthropocene* (Oakland, CA: University of California Press, 2016) for discussions on the Anthropocene.

6   Zhihe Wang, Huili He, and Meijun Fan, "The Ecological Civilization

Debate in China," *Monthly Review* 66, Number 6 (November 2014): 73.

7   Yijie Tang, "Reflective Western Scholars View Traditional Chinese Culture," in *The People's Daily,* February 4, 2005.

8   Wang Jin, "China's Green Laws Are Useless," China Dialogue, September 23, 2010, https://chinadialogue.net.

9   Robert Boyle, *A Free Enquiry into the Vulgarly Received Notion of Nature,* edited by Edward Davis and Michael Hunter (Cambridge: Cambridge University Press, 1996), 57.

10 Alfred North Whitehead, *Adventures of Ideas* (New York: The Free Press, 1933), 272.

11 For further information, see https://pandopopulus.com/.

Question 5

WHAT OTHER MOVEMENTS ARE ALLIED WITH ECOLOGICAL CIVILIZATION?

WE HAVE BEEN EXPLORING the emerging vision of an ecological civilization—a sustainable form of civilization that can work for the well-being of all people and the planet, and can do so over the long term. We are in the midst of a terrifying and exciting time in planetary history. The world that will be inhabited by the next 30 generations (or more) depends on which way humanity turns at this crossroads—a decision that will be made, one way or the other, within the next few decades. Like the 95 theses by Martin Luther that launched the Protestant Reformation, the vision of an ecological society can be the beginning of a global movement toward revolutionary change.

A Vision with Many Names

These are undoubtedly bold claims, and certainly some skeptics will resist. After all, they may argue, why would one need a book titled *What is an Ecological Civilization?* if it is already clear that we are in the midst of a global reformation? In truth, it isn't until one realizes that the vision of an ecological civilization has been taking root in many different communities, under many different names, that the scale of this "great transition" gradually becomes clear.

WHAT IS ECOLOGICAL CIVILIZATION?

Almost month by month more leaders are coming to the realization that fundamental changes are required if we are to remedy the exploitation of the earth, which includes the exploitation of ecosystems, animals, and the poorest members of the human population. One hears stronger and stronger calls to action from farmers, educators, economists, politicians, business leaders, and heads of nonprofits—actions that offer the hope of radically changed social structures that promote the common good. From Pope Francis in Rome to President Xi in China, from the United Nations to the Parliament of the World's Religions, proposals are being made for practices, technologies, and lifestyle changes that seek to bring about a different kind of future.

But *what* kind of future? Do their sometimes very different terms really refer to the same goal, or are they offering vastly different visions for a future society? Where is there common ground? In the next few pages we consider some of the main concepts that are being used: "integral ecology," the "new story," "unity in diversity," "constructive postmodernism," the "ecozoic era," the Japanese notion of "Yoko civilization," "organic Marxism," "process philosophy," an "open and relational worldview," and the "philosophy of organism."

The differences are real. Some of them emphasize changing the paradigms that people use to interpret the world as a whole. For example, one might switch from materialism to an organic paradigm that construes reality as relationships rather than objects, so that the universe is seen as a living, relational organism. For some, core values need to be rethought. Here the emphasis may fall on metaphysical beliefs, or religious motivations, or spiritual practices. For some, new voices need to become the guides for humanity—the voices of women, indigenous peoples, the poor, or reformers in the Global South. For some, what are needed are systemic political or economic changes—concrete structural changes that restrict the power of capital and support socialist forms of organization that meet the core needs of all citizens.

Still, beneath all these options one discerns the outlines of a common vision. It's the vision of moving beyond the broken civilization that is driving economic globalization today—the vision of a radically new mode of human life on this planet.

Question #5

Integral Ecology as Ecological Civilization

Perhaps the most prominent figure arguing for fundamental societal change is the charismatic Pope Francis. It's no coincidence that Jorge Mario Bergoglio chose the name Francis to signify his affinity to St. Francis of Assisi, the patron saint of animals and the environment. As the leader of the Roman Catholic church, Francis has a following of some 1.2 billion Catholics throughout the world and is highly respected by secularists and other religious traditions alike. The importance of his selecting this papal name was brought home in 2015, when his papal encyclical *Laudato Si'* ("Praise be to You") confirmed his desire to emulate the spirit of his namesake.

In his groundbreaking encyclical, Pope Francis inspired the world with his unwavering commitment to truth, to rigorous scientific research of the highest quality, and to core spiritual values found in many of the different religious traditions around the world. (It's no coincidence that *Laudato Si'* is addressed not only to Catholics, but also to humankind as a whole.) The encyclical calls readers to live in ways that transform and empower all of humanity to serve the global common good. Its central theme, "Care for Our Common Home," includes not only environmental justice, but is also essentially connected to social and economic justice. Francis notes, for example, that "the poverty and austerity of Saint Francis were no mere veneer of asceticism, but something much more radical: a refusal to turn reality into an object simply to be used and controlled."[1]

What makes the pope's approach inspiring is his formulation of a comprehensive integrative paradigm for human action. As a way to explain this holistic approach to collective well-being, Francis uses the phrase *integral ecology*. Integral ecology is based upon a deep realization of the complex interrelationships that make up the universe, ranging from physical molecules to ecosystems to societal dynamics to global cultures.[2] For example, in striving to understand the environmental predicament, he avoids isolating problems. Instead of reducing them to specific contexts and causes, he digs down deeper to the systemic issues. For example, he writes

> Recognizing the reasons why a given area is polluted requires a study of the workings of society, its economy, its behavior patterns, and the ways it grasps reality. Given the scale of change, it is no longer possible to find a specific, discrete answer for each part of the problem.[3]

This excerpt captures the spirit of integral ecology, as well as the spirit of ecological civilization. Both terms refer to a systematic approach that works to name the underlying causes of the ecological crisis in all its dimensions. Like ecological civilization, Pope Francis' integral ecology draws together a web of interrelated factors: societal, political, cultural, psychological, epistemological, ethical, and spiritual.

Just as the term "civilization" encompasses all the diverse elements that make up contemporary global society, the term "integral" implies a deep interconnectedness that transcends arbitrary boundaries between individual and societal, between human and nonhuman, between the environment and society. Pope Francis argues that "we are faced not with two separate crises, one environmental and the other social, but rather with one complex crisis which is both social and environmental."[4]

The complexity and interrelatedness of the crisis lead Francis to conclude that simple or superficial changes are not radical enough to repair the broken systems of contemporary society; what is truly needed are fundamental changes in cognitive habits and patterns of thought. Here we find his most radical invitation: the call to internally transform our values as individuals and societies. To this point he writes, "Our efforts at education will be inadequate and ineffectual unless we strive to promote a new way of thinking about human beings, life, society and our relationship with nature."[5] Elsewhere he calls this an *ecological spirituality*.

In these comparisons one can see the deep affinity between Pope Francis' call to embrace an integral ecology and the vision of an ecological civilization presented in this book. Both integral ecology and ecological civilization entail fundamental changes, not only in the way that we live, but also in the ways that we think; they constitute a paradigm shift toward a more sustainable and just world. Both integral ecology and ecological civilization push beyond the silos of binary and disciplinary thinking, replacing them with a more holistic worldview. Pope Francis' vision for an integral ecology resonates deeply with the call to leave behind the unsustainable patterns of modern civilization in favor of new ways to organize human existence on this planet.

The EcoCiv Vision as a New Story or Paradigm: David Korten

Like Pope Francis, David Korten has famously called for a global paradigm shift away from the exploitive system of capitalism to a new "living systems" paradigm. In his important 2015 book, *Change the Story, Change the Future: A Living Economy for a Living Earth*,[6] Korten demonstrates how humanity's most urgent problems (i.e., environmental, social, and economic injustices) are systemic problems. And systems problems can often be traced back to faulty paradigms.

According to Korten, the story that now governs our society, "Sacred Money and Markets," has set us on a destructive path. It is a story—a paradigm—that alienates us from nature by treating the earth as nothing more than a resource to be exploited in the pursuit of money. As Korten explains:

> Human well-being is best assessed by indicators of the health and happiness of our children, the strength of our families and communities, the purity of our air and water, the stability and fertility of our soils, and the vitality of our forests and fisheries. In denial of this seemingly obvious reality, we have allowed ourselves to become captive to economic theories and institutions that would have us believe that economic health is best assessed by the growth of a financial indicator, gross domestic product (**GDP**).[7]

In order to combat the injustices ushered in by transnational corporate rule, Korten suggests changing the story whereby we can change our future. We need an alternative story that "reflects the fullness of human knowledge and understanding and provides a guide to action to meet the needs of our time."[8] The new story stands in opposition to the corporate story of money as the sole source of happiness. This narrative recognizes relationships as the source of happiness. It views the world as a living organism instead of as a set financial resources, stressing compassion and the common good rather than greed and individualism. The alternative story—the "Sacred Life and Living Earth" story—represents a fundamental shift in the way we understand ourselves in relation to the living universe. When life rather than money becomes our core value, we begin to see the world, and one another, in a whole new light.

Korten's influential Living Earth story exactly reflects the vision of a sustainable civilization. Korten proposes this new paradigm as a means:

1. to maintain a healthy balance between humans and the generative systems by which the Living Earth continuously renews herself; and

2. to secure for all people the essentials of human health and happiness.[9]

These goals also reflect the mission of the Club of Rome, which describes itself as "an organization of individuals who share a common concern for the future of humanity and strive to make a difference." The Club states that its mission is "to act as a global catalyst for change through the identification and analysis of the crucial problems facing humanity and the communication of such problems to the most important public and private decision makers as well as to the general public."[10] The Club of Rome, of which Korten is a member, has spent decades advocating for a fundamental shift away from systems of exploitation that are designed to benefit the few at the expense of the rest. By taking a global perspective and affirming the interconnectedness of all things, the Club seeks holistic solutions for the long-term benefit of all.[11] This integral method offers interesting parallels with the pope's integral approach to ecology.

Changing the story humanity tells of itself and its values is a means of addressing the deepest problems—not just in the way we live, but also in the ways we think. Any possible paradigm shift toward a more sustainable and just world will have to begin with a radically new story and set of goals. Treating science and policy analyses only as endeavors of experts within the silos of specialized disciplines makes it impossible for leaders to be guided by a larger, more holistic worldview that values the common good.

More recently, Korten himself has recognized the deep affinity between his Living Earth paradigm and the vision of an ecological civilization. In a recent address at the University of Alberta, Korten affirmed that "the task at hand is to navigate a civilizational shift from the imperial civilization of the past 5,000 years, to the ecological civilization on which our future depends." He continues, "There is no blueprint or easy set of prescriptions for the turning to an Ecological Civilization. I can, and will, however, offer some fundamental principles and guidelines—starting with two foundational design priorities: We must value life over money. And the relationship of community over the isolated individual." [12]

Like integral ecology and ecological civilization, and building on the legacy of the Club of Rome, Korten offers an approach that seeks transformation at the most fundamental levels: a shift in values that prioritizes life over money and promotes the common good. Korten's demands for reform, going back to his early work, *When Corporations Rule the World*, break so radically with the current world order that they can only be interpreted as a call for civilizational change.[13]

Ecozoic Era: Cosmological Perspective for Ecological Civilization

In 1992 Thomas Berry and Brian Swimme published their epic book, *The Universe Story: From the Primordial Flaring Forth to the Ecozoic Era—A Celebration of the Unfolding of the Cosmos*. It was in that book that they coined the term "Ecozoic era" as a way to describe an era in which "humans live in a mutually enhancing relationship with Earth and the Earth community."[14]

Much like Korten, who urges us to change our story, Berry and Swimme have encouraged a reframing of human civilization in light of cosmic history itself, a story they call "the universe story." Consistent with integral ecology, the Living Earth story, and ecological civilization, the ecozoic era also requires a fundamental transformation of our systems and paradigms, of the ways that human civilization has been organized up to this point. In describing the contemporary context, Thomas Berry writes,

> Our present system, based on the plundering of the Earth's resources, is certainly coming to an end. It cannot continue. The industrial world on a global scale, as it functions presently, can be considered definitively bankrupt. . . . The Earth cannot sustain such an industrial system or its devastating technologies. . . . The impact of our present technologies is beyond what the Earth can endure.
>
> The changes presently taking place in human and earthly affairs are beyond any parallel with historical change or cultural modification as these have occurred in the past. This is not like the transition from the classical period to the medieval period or from the medieval to the modern period. These changes reach far beyond the civilizational process, beyond even the human process, into the biosystems and even the geological structures of the Earth itself.[15]

Berry's hard-hitting observations are quite similar to the starting point advocated by John Cobb, who set the theme for the 2015 Ecological Civilization conference. Cobb writes,

> We live in the ending of an age. The age that ends may be just the modern period. . . . The amazing achievements of modernity make it possible, even likely, that its end will also be the end of civilization, of many species, or even of the human species. At the same time, there are many new beginnings that give promise of an ecological civilization. There have been numerous conferences devoted to finding a way to sustain civilization, and many promising practical responses have been proposed. There have also been conferences showing how the modern world has led us to the brink of catastrophe and calling for changes in basic attitudes and orientations. We support and celebrate much that has been accomplished.[16]

Clearly Cobb and Berry have similar views regarding the threat that the planet now faces. Both also trace the current crisis back to the "father of modern philosophy," René Descartes. As we have seen above, the organic worldview that underlies ecological thinking contrasts sharply with the mechanistic worldview promoted by Descartes and many modern philosophers. Likewise, Thomas Berry seeks to clearly distinguish the ecozoic era from the Cartesian model:

> Descartes, we might say, killed the Earth and all its living beings. For him the natural world was mechanism. There was no possibility of entering into a communion relationship. Western humans became autistic in relation to the surrounding world. There could be no communion with the birds or animals or plants, because these were all mechanical contrivances. The real value of things was reduced to their economic value. A destructive anthropocentrism came into being. *This situation can be remedied only by a new mode of mutual presence between the human and the natural world*, with its plants and animals of both the sea and the land.[17]

In this passage one also hears the resonance with David Korten's Living Earth story; both call for shifting core values away from individualism and commercialism, and replacing them with values that lead to the flourishing of all life. A sustainable way of life at the global level will require fundamental reforms in both thought and action. Minor changes in behavior are not enough to overcome the destructive modern paradigm of exploitation, which was the underlying story that defined the era that is now ending. Instead, Swimme and Berry argue that a radical reorientation of humanity's relationship to the planet is required. As Berry notes, "The Ecozoic Era can be brought into being only by the integral life community itself. . . . For this to emerge there are special conditions required on the part of the human, for although this era cannot be an anthropocentric life period, it can come into being only under certain conditions that dominantly concern human understanding, choice, and action."[18]

Among the conditions that are necessary in order to usher in a new ecozoic era, Berry lists three:

1. an understanding that the universe is a communion of subjects, not a collection of objects;

2. a realization that the Earth exists, and can survive, only in its integral functioning; it cannot survive in fragments any more than any organism can survive in fragments;

3. a recognition that the Earth is a one-time endowment; we must reasonably suppose that the Earth is subject to irreversible damage in the major patterns of its functioning and even to distortions in its possibilities of development.

All of these conditions reflect the same vision of transitioning to an ecological civilization. What's striking about them is they are not strictly about changing our actions, such as passing policies to support renewable energy and regenerative agriculture. Instead, they focus on fostering new recognitions and realizations, which serve as the foundations for a new paradigm. New modes of mutual presence between humans and the natural world can then follow.

While not identical, the general vision that Thomas Berry calls an ecozoic era overlaps significantly with what David Korten calls the Living Earth story, what Pope Francis calls an integral ecology, and what we call ecological civilization.

Yoko Civilization: A Spiritual Vision for Moving Beyond Materialism

Kōtama Okada founded Sukyo Mahikari in Japan in 1959. This global movement emphasizes shared principles among the various religions (*sukyo*) and spiritual experiences of light and healing (*mahikari*). Along with practices, Okada also advocated for new forms of social organization. A Yoko civilization—positive or bright (*yo*) + light (*ko*)— is one that shifts away from "a material-centered way, which pursues material wealth by exploiting nature," to "a spiritually oriented way."[19] Humans will thrive only when we stop centering our attention on personal gain and wealth and learn to live with each other and with nature in ways that are "spirit-centered."

Consistent with the proposals for change that we have already examined, the vision of a Yoko civilization is holistic, integrative, and systemic in nature. Prioritizing the common good, it works to overcome conflict and disintegration and to realize peace through a radical reorientation of humanity's relation to the earth. Spiritual reorientation involves a profound transformation of the most fundamental aspects of who we are. But, Okada maintains, this reorientation is not merely internal; ultimately it's about civilizational change on a global scale. Only when individuals and societies break with the exploitive materialism that underlies international economics and politics can we begin to realize a harmony between humans and the environment.

The Earth Charter

Like Pope Francis' *Laudato Si'*, the Earth Charter is an example of a visionary document that attempts to provide a holistic framework for civilizational change. After the World Commission on Environment and Development (known as "the Brundtland Commission") in 1987, there was a call for drafting a new charter that would serve to guide global transition toward sustainable development. In March of 2000, after years of international collaboration on drafts and revisions that incorporated the input of people from around the world, the Earth Charter Commission reached consensus on the document known as the Earth Charter.

While the Charter has not received the level of international support necessary for global implementation, it does represent "a global consensus statement" on the meaning of sustainability, the challenge and vision of sustainable development, and the principles by which sustainable development is to be achieved.[20]

The Earth Charter represents as "an ethical framework for building a just, sustainable, and peaceful global society in the 21st century." Like the vision of ecological civilization, the Earth Charter emphasizes the principle of interconnectedness and concern for the common good (human and beyond). The Earth Charter Initiative also describes the Charter as "a vision of hope and a call to action," which seeks to "inspire in all people a new sense of global interdependence and shared responsibility for the well-being of the whole human family, the greater community of life, and future generations."[21]

WHAT IS ECOLOGICAL CIVILIZATION?

The Charter includes a robust list of 16 core principles, with more than 60 sub-principles, organized under four broad categories. These core commitments and values cover a range of social, political, spiritual, and environmental issues. The Charter begins with the four broad commitments regarding respect and care for the community of life:

1. Respect Earth and life in all its diversity.
    a. Recognize that all beings are interdependent and every form of life has value regardless of its worth to human beings.
    b. Affirm faith in the inherent dignity of all human beings and in the intellectual, artistic, ethical, and spiritual potential of humanity.
2. Care for the community of life with understanding, compassion, and love.
    c. Accept that with the right to own, manage, and use natural resources comes the duty to prevent environmental harm and to protect the rights of people.
    d. Affirm that with increased freedom, knowledge, and power comes increased responsibility to promote the common good.
3. Build democratic societies that are just, participatory, sustainable, and peaceful.
    e. Ensure that communities at all levels guarantee human rights and fundamental freedoms and provide everyone an opportunity to realize his or her full potential.
    f. Promote social and economic justice, enabling all to achieve a secure and meaningful livelihood that is ecologically responsible.
4. Secure Earth's bounty and beauty for present and future generations.

g. Recognize that the freedom of action of each generation is qualified by the needs of future generations.

h. Transmit to future generations values, traditions, and institutions that support the long-term flourishing of Earth's human and ecological communities.

As you can see, these principles—which only represent a portion of "the Earth Charter Global Movement"[22]—closely resemble the commitments of the movement toward ecological civilization. In addition to the principles of interdependence, the value of all life, and a framework that combines the social and environmental, the Earth Charter movement and the Seizing an Alternative movement share the conviction that transitioning to an ecological civilization will require a change of mind *and* heart, until the flourishing of life takes priority over economic growth.

From ecological protection to the eradication of poverty, from economic equality to peace and justice, the Earth Charter provides a set of guidelines for a new kind of society—a new form of civilization—grounded in an inclusive ethical framework and an integral worldview that is oriented toward long-term sustainability and the common good.

Constructive Postmodernism and Ecological Civilization

Nowhere in the world has the notion of an ecological civilization received more attention than in the People's Republic of China. In 2012, the Communist Party of China included the goal of achieving an ecological civilization in its constitution, and also featured the goal in its five-year plan.[23] In announcing the next five-year plan in 2017, President Xi again stressed the same notion. No one is under any illusions about the difficulty of the task. But even to acknowledge the importance of civilizational change puts China far ahead of the United States.

In researching what an ecological civilization might entail, Chinese scholars have identified a number of related concepts. As we will see below, process-related categories have played a major role in their thought. A more environmentally friendly approach to Marxism, which is enshrined in the Chinese Constitution, has also been developed under the heading of "organic Marxism." Central to Chinese attempts to flesh out the implications of ecological civilization, at least in the early years, was the notion of *constructive postmodernism*.

The term "postmodern" has often been used to articulate alternatives to modernism and was popularized by French philosophers such Jean-François Lyotard, Jacques Derrida, and Michel Foucault. French postmodernism is often described as "deconstructive," since it seeks to deconstruct modern assumptions, values, and ideas. The label "constructive postmodernism" was coined by the American process philosopher David Ray Griffin in 1989 as a way to offer a positive alternative to deconstructive postmodernism. As John Cobb notes, "The term 'constructive' is used to contrast with 'deconstructive' to emphasize that constructive post-modernism is proposing a positive alternative to the modern world. This does not mean that it opposes the work of deconstructing many features of modernity. The point is that critique and rejection should be accompanied by proposals for reconstruction." [24]

Griffin went on to edit an important series on Constructive Postmodernism with the State University of New York Press. Of all the books in this series, however, it was his volume *The Reenchantment of Science* that has had the greatest international influence. When it was translated and published in China it quickly spawned an influential movement, which caught the attention not only of university professors but also of leading figures in the government.

The reason for the influence of constructive postmodernism in China is significant. In the wake of the Cultural Revolution, China was in search for a new way forward. Was it possible to modernize a largely agrarian, developing economy without falling into the errors exemplified by the West, including colonization and environmental degradation? Constructive postmodernism caught on as a way to highlight this goal. By contrast, the deconstructive works of the French philosophers seemed only to criticize Western modernism without offering a clear alternative. To Chinese leaders, Griffin's constructive version of postmodernism seemed to offer the solution. Tang Yijie, a well-known professor at Peking University, describes the history:

Question #5

In the 1990s, there emerged in China's ideological and cultural circles two ideological trends opposing the concept of "modernism." One trend is "postmodernism," an idea originating in the West and aiming to deconstruct "modernity." In the early 1980s, "postmodernism" had already come to China, but it had little impact at the time. By the 1990s, however, Chinese scholars were suddenly showing it great interest. . . . At the turn of the 21st century, "constructive postmodernism," a concept based on process philosophy, proposed integrating the positive elements of the first Enlightenment with postmodernism and thus called for a "second enlightenment."[25]

As Tang explains, constructive postmodernism would eventually become one of the major schools of thought in China, embraced for its ability to synthesize the best of modernity with the best of traditional Chinese culture, in order to formulate a positive vision of harmony with one another and the planet.[26]

Griffin describes constructive postmodernism as the vision of a "society organized for the good of the planet as a whole."[27] Like the vision of an ecozoic era described above, constructive postmodernism proposes a radical reorientation of our most fundamental understanding of the world, an alternative cosmology that challenges mechanistic and dualistic thinking in science and philosophy. Constructive postmodernism proposes a replacement of modern dualism and reductionism with a postmodern ecological and organismic paradigm.

From dualism to holism, from mechanism to organism, from individualism to the common good, it's evident that constructive postmodernism overlaps significantly with the vision of an ecological civilization. In the view of many Chinese leaders and scholars, it is an indispensable step in that direction.

Organic Marxism: A Political Vision for Ecological Civilization in China . . . and Beyond

Closely related to the vision of constructive postmodernism and ecological civilization is the more recent proposal of a more "organic" and environmentally friendly form of Marxism. Over the last few years, the vision of a truly organic Marxism has become deeply intertwined with the Chinese discussion of ecological civilization and constructive postmodernism.

Dr. Zhihe Wang speaks of the positive reception of constructive postmodernism in connection with Marxism, stating that:

> Constructive Postmodernism has deep convergences with Chinese Marxism, such as putting emphasis on process, taking an organic stance, having a strong consciousness of social responsibility, caring for the poor, defending justice, and pursuing the common good. I argue that it is these deep convergences that make some open Chinese Marxists enthusiastic about Constructive Postmodernism.[28]

This apparent affinity between Chinese Marxism and constructive postmodernism is further elucidated in the book by Philip Clayton and Justin Heinzekehr, *Organic Marxism: An Alternative to Capitalism and Ecological Catastrophe*. Clayton and Heinzekehr draw connections between economic and political systems and the climate crisis. Their "manifesto" for a sustainable society is built on a number of core assumptions and observations:

1. It is urgent that people around the planet recognize the situation that the human race now faces and take steps to implement the solutions.
2. Capitalism as a social and economic system has created massive injustices and has devastated the global environment.
3. There are real alternatives to unrestrained capitalism.
4. One alternative is a hybrid system that limits market forces and supports social communities structured for the common good.
5. The power of corporations and wealthy individuals—what the Occupy movement called the 1%—comes at the expense of the vast majority of the world's population, the 99%. Capitalist policies favor the 1%; socialist policies prioritize the interests of the 99%. This is a dynamic that Karl Marx studied in the 19th

century and that Thomas Piketty's recent analysis of economic trends across the 20th century overwhelmingly confirms.[29]

6. It is now virtually inevitable that global climate change will produce social and economic collapse on many parts of our planet.

7. Out of the dust of that collapse, a new ecological civilization can arise.

8. It is far better for humans and for the planet if we act now rather than waiting for the full force of the calamity to strike.

These priorities of organic Marxism are also represented in different ways by the other schools of thought covered in this chapter. Given its specific criticisms of capitalism and exploitive economic systems, organic Marxism is most closely linked with the vision of David Korten and the Club of Rome.

Almost by definition, value is what we aim to increase. Organic Marxism calls for a reorientation of values, not just from individualism toward community, but also from anthropocentrism toward planetary flourishing. Additionally, how we define success matters. As long as success is thought of in terms of power and money, fame and fortune, big houses and fancy cars, societies will continue to move toward environmental collapse, and socio-economic inequality will continue to increase. Profit- and consumer-based notions of "success" tend toward selfishness and alienation from nature. At its core, organic Marxism is a call toward a paradigm shift that prioritizes the common good.

Organic Marxism is ecological in several senses. In opposition to mechanistic worldviews, it promotes a worldview based on organic relations between humans and with nature. Healthy ecosystems are based on a balance between multiple agents. All parts are interconnected, and out of the totality of the parts, systems emerge that in turn sustain their parts. Healthy societies and economic systems function in the same way. The ecological model promotes a shift in value, prioritizing organic human-nature systems instead of extractive and hierarchical systems. Without this shift, no civilization can be sustainable in the long run.

Philosophy of Organism: A Relational Worldview for the Common Good

In his seminal work, *Process and Reality*, mathematician and philosopher Alfred North Whitehead outlines what he calls a "philosophy of organism," giving birth to what would later be called "process philosophy." Given that David Ray Griffin, John B. Cobb, Jr., and several of the proponents of ecological civilization have been deeply influenced by Whitehead's thought, it should be no surprise that process philosophy is part of the roadmap toward an ecological civilization. Process thinkers are proponents of "an open and relational worldview for the common good."[30]

At its core, Whitehead's philosophy of organism provides an alternative way of understanding the nature of reality. The philosophy of organism transitions from a substance-based ontology to an event-based ontology, from a mechanistic framework to an organismic framework, from a dualistic worldview to a holistic worldview, from being to becoming. To understand the world in these new terms is to promote a fundamental shift in values.

Much like Thomas Berry's conditions to usher in a new ecozoic era, Whitehead's philosophy of organism shifts from a collection of objects to a communion of subjects, placing priority on the integrated whole rather than on fragmented parts. Both Berry and Whitehead recognize that reality is in the end better understood as living systems within systems within systems, rather than as non-living matter that can be understood and calculated through physical equations alone.

Developed in contrast to the mechanistic dualism of Descartes, Whitehead's philosophy of organism offers a framework for understanding the universe as a creative living system in which *all* agents are intrinsically valuable. While not all supporters of the transition toward an ecological civilization are proponents of Whitehead's philosophy of organism, nearly all supporters of Whitehead's philosophy are proponents of the vision of an organic ecological civilization.

The United Nations and the Sustainable Development Goals

While the ice caps melt, sea levels rise, and species go extinct—and in part *because* of the growing climate crisis—our world is characterized by radical economic inequality and systemic social injustices. The problems we face are environmental, social, and deeply systemic. When so many features of globalization are unjust, it is particularly difficult to imagine concrete steps toward change. It's for this reason that the United Nation's Sustainable Development Goals are so significant

The process of drafting "Agenda 2030," the U.N.'s Sustainable Development Goals, involved 193 Member States and a global collaboration with civil society leaders. Agenda 2030 describes the threats to society and the planet:

> Global health threats, more frequent and intense natural disasters, spiraling conflict, violent extremism, terrorism and related humanitarian crises and forced displacement of people threaten to reverse much of the development progress made in recent decades. Natural resource depletion and adverse impacts of environmental degradation, including desertification, drought, land degradation, freshwater scarcity and loss of biodiversity, add to and exacerbate the list of challenges which humanity faces. Climate change is one of the greatest challenges of our time and its adverse impacts undermine the ability of all countries to achieve sustainable development.[31]

The U.N. Sustainable Development Goals (SDGs) identify specific paths for addressing these challenges across the multiple sectors of today's global civilization, focusing specifically on social and environmental issues. One gets a picture of how comprehensive the SDGs are by considering the range of issues that they cover:

1. *Zero Poverty*: End poverty in all its forms everywhere.
2. *Zero Hunger*: End hunger, achieve food security and improved nutrition, and promote sustainable agriculture.
3. *Good Health & Well-Being*: Ensure healthy lives and promote well-being for all at all ages.
4. *Quality Education*: Ensure inclusive and equitable quality education and promote lifelong learning opportunities for all.

5. *Gender Equality*: Achieve gender equality and empower all women and girls.

6. *Clean Water & Sanitation*: Ensure availability and sustainable management of water and sanitation for all.

7. *Affordable & Clean Energy*: Ensure access to affordable, reliable, sustainable, and modern energy for all.

8. *Decent Work & Economic Growth*: Promote sustained, inclusive and sustainable economic growth, full and productive employment, and decent work for all.

9. *Industry, Innovation, & Infrastructure*: Build resilient infrastructure, promote inclusive and sustainable industrialization, and foster innovation.

10. *Reduced Inequalities*: Reduce inequality within and among countries.

11. *Sustainable Cities & Communities*: Make cities and human settlements inclusive, safe, resilient, and sustainable.

12. *Responsible Consumption & Production*: Ensure sustainable consumption and production patterns.

13. *Climate Action*: Take urgent action to combat climate change and its impacts.

14. *Life Below Water*: Conserve and sustainably use the oceans, seas, and marine resources for sustainable development.

15. *Life On Land*: Protect, restore and promote sustainable use of terrestrial ecosystems, sustainably manage forests, combat desertification, and halt and reverse land degradation and biodiversity loss.

16. *Peace, Justice, & Strong Institutions*: Promote peaceful and inclusive societies for sustainable development, provide access to justice for all, and build effective, accountable, and inclusive institutions at all levels.

17. *Partnerships For The Goals*: Strengthen the means of implementation and revitalize the Global Partnership for Sustainable Development.

Undergirding these 17 overarching goals are 169 specific targets and 230 individual indicators, which serve as roadmaps for action and criteria for success. Together, the SDGs form a vision for the kind of world that 193 countries seek to build. As the U.N. states:

> We envisage a world in which every country enjoys sustained, inclusive and sustainable economic growth and decent work for all. A world in which consumption and production patterns and use of all natural resources—from air to land, from rivers, lakes and aquifers to oceans and seas—are sustainable. One in which democracy, good governance and the rule of law, as well as an enabling environment at national and international levels, are essential for sustainable development, including sustained and inclusive economic growth, social development, environmental protection and the eradication of poverty and hunger.[32]

Unfortunately, there are also serious problems with the SDG approach. The biggest problem with current thinking about sustainable development is that it is often framed in ways that suggest that the goals can be met without radical systemic change. Many of the targets and indicators that are used to measure success in achieving the SDGs turn on the amount of financial investment and the GDP of individual nations. But increasing someone's wages from $1.25 per day to $2 per day is an inadequate measure of how well poverty and human well-being are being addressed. Increasing GDP is not enough to help resolve the problem of income inequality when eight individuals possess as much as half of the world's population.[33]

Likewise, efficient recycling isn't going to create sustainable cities if we don't reduce consumption. The targets and indicators underlying the goals for sustainable development reveal a continuing reliance on contemporary economic assumptions to the extent that they still presuppose a system that exploits people and nature for profit and power.

Still, there are many working with theSDGs who openly call for a shift away from business as usual. The SDGs are caught in a wrestling match between two different understandings of the term "sustainability": sustaining the current economic system while "greenifying" it where possible, or rethinking it from bottom up (and top down). The goal of sustainable development, *taken in the broadest sense*, can represent a call for a holistic and integral framework. We cannot adequately solve the hunger problem without addressing poverty and food insecurity; we cannot significantly decrease food insecurity and poverty without addressing climate change and economics; and we cannot slow climate change without addressing systems of capitalism, unlimited growth, and the values that prioritize short-term convenience over the long-term well-being of people and the planet.

Neither side doubts the severity of the global climate crisis and its effects across geographies and societies. Transformation requires us to consider development in the context of systemic change. Admittedly, when one faces a global problem, it is often helpful to "chunk it" into individual goals in order to make it more manageable. But the discussion around the SDGs has tended to divide them into silos, resulting in actions that are more isolated and fragmented and thus (paradoxically) less effective in addressing the fundamental causes. This contrasts with more integral and holistic approaches to global sustainability; for example, the Vatican's Dicastery for Promoting Human Development under Cardinal Turkson, which advocates for systemic conclusions in ways similar to the ecological civilization movement.

What Unifies These Movements?

Whether we call it integral ecology, Yoko civilization, sustainable development, organic Marxism, constructive postmodernism, an ecozoic era, a Living Earth story, or an ecological civilization, a common vision appears to be emerging around the globe. It is a vision for a radically different future in which the systems of society—economics and politics, systems of production, consumption and agriculture—are redesigned in light of planetary limits. The vision emerges from the death of the modern *expand, conquer, and consume* mentality and tells a new story of harmony with a living Earth: *contract, cooperate, and cultivate*.

While all of these distinct movements share core values, philosophies, and even hopes for the future, perhaps the single greatest unifying feature is the call to fundamental systemic change across all areas

Question #5

of society. Nationalism, war, political corruption, economic disparities, overpopulation, crop failures, drought, famine—these are all individual strands in a tangled ball of string. It does not look likely that governments have the will (or the time remaining) to untangle the strings and make repairs. Apparently, whether governments intend it or not, broader systematic changes are on the way. Therein lies the fear . . . but also the hope.

Imagine a fisherman on a boat. The boat is filling with water. He faces a critical problem; if he doesn't solve it, the boat will sink. So the fisherman takes a bucket and starts throwing water overboard, over and over again, until his arms are so tired that he can no longer continue. Even though he threw hundreds of buckets of water out of the boat, he's still going to sink. But there's another option: fix the hole in the boat! This call to address the underlying causes of our world's most critical problems is what unites the parallel movements explored in these pages. The problems cannot be neatly divided into social, economic, political, or environmental. As Pope Francis notes, we have "one complex crisis which is both social and environmental." To focus on environmental issues without considering the social, or vice versa, is to fail to address the root causes of the crisis.

To clarify: it remains crucial to take steps to alleviate the symptoms as well. Like the fisherman bailing water, it's necessary to decrease carbon emissions, turn to renewable energy sources, reform education, and limit the injustices in economic systems. Yet, if we pay attention only to the symptoms and not also to the underlying causes, long-term success will evade our grasp. This holistic call toward fundamental transformation at all levels of society and across all sectors is the mark of the ecological civilization movement.

At this point, some readers may be concerned about how ambitious it is to strive for civilizational change to a more ecological age. Fundamental transformation involving all sectors and levels of society is no minor task. Is it even possible, and, if so, where does one start? These are the questions to which we turn next.

Endnotes

1. Francis, *Laudato Si': Encyclical Letter of Pope Francis on Care for Our Common Home* (Vatican City: Libreria Editrice Vaticana. Promulgated 24 May, 2015), http://w2.vatican.va/content/ francesco/en/encyclicals/documents/papa-francesco_20150524_enciclica-laudato-si.html., §11.
2. *Laudato Si'*, §138: "It cannot be emphasized enough how everything is interconnected."
3. *Laudato Si'*, §139.
4. *Laudato Si'*, §139.
5. *Laudato Si'*, §215.
6. David C. Korten, *Change the Story, Change the Future: A Living Economy for a Living Earth* (Oakland, CA: Berrett-Koehler Publishers, Inc., 2015).
7. David C. Korten, "Birthing an Ecological Civilization: Overview," published April 14, 2017, http://davidkorten.org/birthing-an-ecological-civilization/.
8. From the back cover of David C. Korten's *Change the Story, Change the Future: A Living Economy for a Living Earth* (Oakland, CA: Berrett-Koehler Publishers, Inc., 2015).
9. Korten, "Birthing an Ecological Civilization."
10. "About Us," Internet Archive, archived December 19, 2008, para. 1, https://web.archive.org/web/20081219222403/http://www.cluboframe.org/eng/about/3/
11. Keith Suter, "The Club of Rome: The Global Conscience," *Contemporary Review* 275, no. 1602 (July 1999): 1–5.
12. David Korten, "A Living Earth Economy for an Ecological Civilization," http://ecociv.org/a-living-earth-economy-for-an-ecological-civilization-david-korten/.
13. David C. Korten, *When Corporations Rule the World*, 3rd Edition (The Living Economies Forum, 2015)
14. See Allysyn Kiplinger's article on Thomas Berry, "What Does

Question #5

Ecozoic Mean," *The Ecozoic Times,* https://ecozoictimes.com/what-is-the-ecozoic/what-does-ecozoic-mean/.

15 Thomas Berry, "The Ecozoic Era," ed. Hildegarde Hannum, The E. F. Schumacher Lecture, https://centerforneweconomics.org/publications/the-ecozoic-era/, .

16 John B Cobb, Jr., "Why This Conference?," http://whitehead2015.ctr4process.org/about/why-this-conference/.

17 Berry, "The Ecozoic Era," italics added.

18 Berry, "The Ecozoic Era."

19 "Yoko Civilization International Conference," Yoko Civilization Research Institute, http://www.ycri.jp/e/.

20 "The Earth Charter," The Earth Charter Initiative, http://earthcharter.org/discover/the-earth-charter/.

21 "Earth Charter Around the World," The Earth Charter Initiative, http://earthcharter.org/.

22 "The Movement: The Earth Charter Global Movement," The Earth Charter Initiative, http://earthcharter.org/act/.

23 Zhihe Wang, Huili He, and Meijun Fan, "The Ecological Civilization Debate in China: The Role of Ecological Marxism and Constructive Post-modernism—Beyond the Predicament of Legislation," https://monthlyreview.org/2014/11/01/the-ecological-civilization-debate-in-china/.

24 John B. Cobb, Jr., "Constructive Postmodernism," ed. Ted Brock and Winnie Brock, http://www.religion-online.org/article/constructive-postmodernism/.

25 Tang Yijie, "Confucianism & Constructive Postmodernism," in *Comparative Studies of China and the West* 1 (2013): 11.

26 Yijie, "Confucianism & Constructive Postmodernism," 11.

27 David Ray Griffin, *The Reenchantment of Science: Postmodern Proposals* (Albany, NY: State University of New York Press, 1988), xiii.

28 Philip Clayton and Justin Heinzekehr, *Organic Marxism: An*

Alternative to Capitalism and Ecological Catastrophe (Claremont, CA: Process Century Press, 2014), 12.

29 See Thomas Piketty, *Capital in the Twenty-First Century* [*Le capital au XXI siède*], trans. Arthur Goldhammer (Cambridge, MA: Harvard University Press, [2013] 2014).

30 See "About The Center for Process Studies," https://ctr4process.org/about/.

31 See the United Nations document, "Transforming Our World: The 2030 Agenda for Sustainable Development," paragraph 14, https://sustainabledevelopment.un.org/post2015/transformingourworld.

32 See the United Nations document, "Transforming Our world: The 2030 Agenda for Sustainable Development," paragraph 9, https://sustainabledevelopment.un.org/post2015/transformingourworld.

33 See the data compiled and published by Oxfam, https://www.oxfam.org/en/pressroom/pressreleases/2017-01-16/just-8-men-own-same-wealth-half-world.]

QUESTION 6

HOW DOES ONE BEGIN BUILDING AN
ECOLOGICAL CIVILIZATION?

If you've read this far, we hope you find the vision of ecological civilization compelling. We hope you're onboard with an ecological worldview, and the need to structure society around the awareness that human beings are one species among others, and that all are interdependent. We hope that you're ready to roll-up your sleeves and get to work developing new forms of sustainable and equitable human communities designed to promote the overall well-being of people and the planet for the long-term. If so, then this "how to" chapter is for you.

How do we begin building an ecological civilization? Asking how one "begins" often implies that one step precedes all others. For some, we need to begin with ideas and beliefs—a vision of ecological civilization that can shape our actions. For others, we need to begin with a set of ecological practices. No doubt, there is a difference between knowing the path and walking the path. As we see it, how one "begins" building an ecological civilization is not a matter of ideas *versus* actions, but a cycle of mutual influence whereby ideas inform actions, and actions inform ideas, and both are constantly at work in the world. We make this point to be clear that what follows is not an attempt to prescribe a step-by-step manual for "saving the world." The very complex nature of the global crisis is such that it requires a plurality of approaches for a variety of diverse contexts. There are no one-size-fits-all solutions. The path toward building an ecological civilization in China will probably be different than the path in Korea, which will be different than the path in Germany, which will be different than the path in the United States, and so on. Therefore, instead of prescribing "the" way one must begin, we will describe some possible ways forward based on how civilization changes tend to (and are already beginning to) occur.

Asking Deeper Questions

One important way that civilizational changes occur is through fundamental paradigm shifts. As we've already discussed in previous chapters, paradigm shifts typically involve a change in worldview—the core ideas that people use to help them understand and interpret the world (see Question Two). There are many ways in which changes in worldview come about. Perhaps the most common cause is that people begin asking new questions in response to changes in the natural and social worlds around them. Asking a new set of questions helps people to think in new ways; in the case of men and women today, it allows us to move closer to understanding the underlying causes of the planet's most pressing problems.

What we've been exploring in this book is civilizational-level change. Because ecological civilization is so fundamentally different from the modern industrial civilizational of exploitation and empire, building an ecological civilization requires a paradigm shift. Such a shift involves a fundamental change to our systems and structures of society as well a fundamental change in the way we think. New ways of thinking invite us to ask new questions. And asking better questions is a critical step toward building an ecological civilization.

Consider the idea that humans are born of and nurtured by a living Earth.[1] In a way, this idea is quite obvious. Humans live on Earth. We're born here. We live here. We die here. This is our home, and we depend upon the Earth for our survival. The less obvious insight (at least for many modern Western philosophers) is that Earth is a living system. Too often we see the world as a collection of objects to be owned, rather than a collection of subjects to be known. What's significant about recognizing that we are born of and nurtured by a *living* Earth? For one, it means that Earth, like all living things, can die. Now, what happens when the biosphere we depend upon for our survival dies? This question naturally follows from the Living Earth worldview. Additional questions can now follow. Is Earth in danger of dying? If so, why? What are the causes of environmental distress or catastrophe? And could we address these underlying causes by changing the way we structure human civilization? You get the picture.

Another fundamental insight from this recognition is that we humans are part of, not separate from, the natural world. As discussed in Question One, and in John Cobb's introduction at the beginning of this volume, "civilization" has been largely characterized by the human ability to manipulate the environment for our own benefit and convenience. When civilizations are structured around human interests divorced from the overall well-being of the living Earth on which we depend, we consume, destroy, and exploit in unsustainable and unjust ways. But what would it look like to build a civilization around the recognition that humans are part of, and dependent upon, a living Earth? How would this change the way we design our cities? How would this change the way we produce and consume food? How would this change the way we educate our children? Beginning with a subtle but profound shift in the way we think, questions emerge with the potential to set us on a new course—toward building an ecological civilization.

One of the most powerful questions we can ask is "why?" Why do eight men possess as much wealth as half of the world's population? Why is topsoil being eroded? Why are global temperatures rising? Why are species going extinct? Each answer provided can be followed by an additional "why?" Deeper and deeper, the "why" questioning exercise takes us closer to understanding the underlying causes of our most serious problems. Through this process, two things become apparent.

First is the recognition that the world's major problems are all interconnected. The global crisis is not neatly divided into separate problems, some social and some environmental. As Pope Francis notes, we have "one complex crisis which is both social and environmental."[2] To focus on environmental issues without considering the social, or the social without the environmental, is a failure to grasp the true nature of the crisis. Consider Agenda 2030, the United Nations Sustainable Development Goals, discussed in the previous chapter. Each of the 17 Sustainable Development Goals (zero poverty, zero hunger, good health, quality education, etc.) is presented as a separate goal, with separate strategies for success. But we can't adequately solve the hunger problem without addressing poverty and food insecurity. And we can't address food insecurity and poverty without addressing climate change and economics, which can't be adequately addressed without dealing with our systems of capitalism, unlimited growth, and the societal values that prioritize short-term convenience over long-term well-being of all people and the planet. True, when facing a big problem, it is often helpful to break it down into smaller pieces to make it more manageable. But if we fail to recognize the interconnected nature of our most serious problems, any solutions we propose will remain fragmented and inadequate.

The exercise of asking deeper and deeper questions also leads to the recognition that the true nature of the global climate crisis is structural and systemic. Adequately addressing the ecological crisis requires changing the systems—economic, political, educational, agricultural, etc.—that form the foundations of contemporary society. One of the biggest problems with current thinking about sustainable development is that it is often framed as possible without radical systemic change. Again, consider Agenda 2030. Most of the targets and indicators used to measure success at achieving sustainable development are related to money and GDP. For example, eradicating poverty is one of the goals (Goal #1, actually). But what is poverty? How is poverty related to money? Does increasing someone's wages (for example, from $2 per day to $4 per day) address the underlying causes of poverty produced by an exploitative economic system that proliferates inequality? If the goals for sustainable development don't include addressing underlying structural and systemic causes of our crisis, then we remain committed, albeit perhaps unconsciously, to the current civilizational systems that exploit people and nature for profit and power. Asking the "why" questions is crucial to the quest for new kinds of systems that are indispensible for building an ecological civilization.

While "why" is an important question, it's not the only one worth asking. Consider the following exercise in "asking better questions" with respect to a specific topic: urban life. Starting with a most basic question, "What is a city?", imagine an answer: "A city is a living system." In fact, one might describe a city as a complex interplay of systems of systems—systems of housing, transportation, sanitation, utilities, governance, communication, economies, etc.—all of which are interrelated and always changing. There are two implications we could extrapolate from this line of thinking. First, when it comes to rethinking urban life with the goal of building an ecological civilization, one should consider how the various parts (systems of transportation, economies, governance, etc.) are related to one another in the context of the whole. When addressing the major challenges of urban life, this "relational" or "systems" thinking approach encourages solutions that work at the level of underlying structural problems and solutions. Second, as a dynamic system of systems, every community is different: different people, different climates, different resources, different histories, etc. That means there is no single right way to structure urban life for all people, in all places, for all time. By beginning with a simple question like "What is a city?", we are drawn to new ways of thinking about urban regeneration, urban-rural relations, and more.

Perhaps the next question for urban life is, "How will city dwellers feed themselves in the future?" Everyone needs to eat, but few of us seriously consider the relationship between rural and urban life. Today's metropolitan centers are dependent on a global system of farming and shipping, wholesalers and retailers. What is visible to consumers in wealthy nations are the first grapes of the season, perhaps grown in the Western Cape region of South Africa. What is not visible are the environmental costs—grapes brought to American dinner tables from the other side of the world arrive with a remarkably high carbon footprint—and the human rights abuses. For example, a day's wage for the grape harvester might not be sufficient to buy the bag of grapes in the Seattle Whole Foods store.

In a society structured around ecological principles—and avoiding global climate collapse will *require* us to build on such principles—urban dwellers will need to be fed by food grown in cities and in the bio-regions around them. So: what will it look like when urban life is restructured so that food needs are met regionally? Perhaps we will replace grass lawns and fields with edible crops, or opt for rooftop and community gardens. The question requires us to begin rethinking the very concepts of "urban" and "rural." It takes us even deeper: perhaps we need to reconsider our societal commitment to a meat-and-dairy-centered diet. The sheer amount of land and resources required to farm livestock makes animal-centered diets unsustainable for densely populated urban communities.

Everyone needs to eat. Everyone also needs to breathe. Yet air quality is a significant challenge for most cities, which is a source of major health concerns for city dwellers. Again the question arises: "Why do urban communities have air quality problems?" The first answers are easy: too few trees, and too many cars. Why? Trees were cut down to pave roads so that suburban dwellers can drive to work, as well as for their food and recreation needs. Increasing urban populations and real estate costs then led to vast regions of low-density communities around the world, resulting in even more cars and fewer trees. At this point, the questions become systematic. While we can develop high-rise forests, turn to electric cars, or bike to work, the underlying problem clearly has to do with how cities have been designed, and how urban growth is still being managed, around the world. Perhaps we should ask, "How can we design cities in ways that promote community belonging while decreasing carbon emissions and pollution?" Decades ago Paolo Soleri proposed "arcologies" as useful models for rethinking city planning for healthy urban living and healthy ecosystems, and in recent years major strides have been taken by architects and city planners to redesign cities for sustainability.

Like eating and breathing, another basic need is shelter from the elements. Everyone needs shelter. You might ask, "How should we structure cities so to ensure adequate shelter for everyone?" Growing demand has made much current housing unaffordable for many city dwellers; the market favors building newer units for wealthy buyers, compounding problems of poverty and ghettoization; and city governments strapped for cash cannot afford to fund moderate- and low-income housing, resulting in increased homelessness and urban sprawl. In an ecological civilization, urban life will need to be restructured to provide adequate housing for urban populations, and in ways that minimize negative environmental impacts. The broader question arises: "What are the underlying obstacles to providing adequate shelter for everyone?" And again it causes us to think more deeply about the underlying structures and systems: how should urban real estate markets be regulated? What city design models provide the environmentally best balance between high-density residency and reduced carbon footprint per person? How must regions around cities be zoned in order to decrease urban sprawl and maximize the capacity of a given region to produce food for its urban population? How are human wishes and environmental needs to be balanced? How fundamentally must the relationship between what we now call "urban" and "rural" be rethought?

In this way, the breadth and depth of the questions about urban life increase tenfold, a hundredfold. "Why do cities tend to be both extremely crowded and increasingly isolating?" leads to "How can we restructure urban life so as to encourage healthy relationships and civic engagement?" "Why are crime rates higher in cities?" leads to "How can we restructure urban life so that people feel safe and secure?" "How should local communities relate to other communities?" leads to "How can urban life be restructured so to promote healthy localization with strong self-governing structures that allow people to cooperate with other communities for mutual flourishing?" The list can easily be extended. Asking deeper questions causes us to seek the necessary data, rethink current assumptions, and explore new kinds of solutions. The resulting process makes the concept of an ecological civilization more precise, more realistic, and more useful.

Top-Down: The Importance of Policy for New Systems

Asking better questions about what an ecological civilization requires is important, but we can't stop at asking questions. We need to seek answers and implement solutions. A changed paradigm has to mean changes to one's life and practice. If a great idea does not lead to social transformation, it is *not* a great idea. So, how does social transformation occur?

One important way is change that proceeds from the large-scale to the small-scale. Top-down change involves action at the top of some organizing body, such as a local, region, or federal government, which affects those members under the jurisdiction of that organizing body. When top-down change involves governments, the laws expand or restrict possibilities for its citizens. Consider the example of the California "burn ban," which prohibits residents from burning wood fires when the amounts of fine particulate matter in the air exceed a certain level. What makes this a top-down act is that the decision not to light wood fires is not made by the individual citizen but by the authorities over a given community—the County of Los Angeles, for example, or the state government in Sacramento.

Different countries have different ideas about the extent to which citizens ought to be able to influence top-town decisions. And of course, there are often vast gaps between what influence people think they *ought* to have, what influence they *think* they have, and what degree of influence they *actually* have. Traditionally, Americans have believed that their government should have less top-down power than the citizens of any other country. It's interesting to contrast this American individualism with China, with its tradition of 2,500 years of centralized governments. Confucian values of placing the good of the community over the wishes of the individual, combined with thousands of years of dynasties, have created a cultural tradition of top-down power, which has the strength to effectively implement policies deemed to be in the public interest. This tradition of centralized governance, and citizens' willingness to accept it, has allowed for unprecedented modernization over the last few decades, including the explosion of industry, rapid technological progress, expansion of urban infrastructures, and the extension of phone and internet access to even remote mountain villages.

At its best, it's an environmentalist's dream: in the same years that California has struggled unsuccessfully to build a single high-speed rail link between LA and San Francisco, China has created a 23,600-mile high-speed rail network. In the expansion of ecological infrastructures, and in multiple green industries such as solar panel production, China has become the global leader. No other nation has built the goal of moving toward an ecological civilization into the platform of the governing party. On the other hand, top-down change is not always gentle. If the Central Government wants to build a train line where a farming village now stands, the villagers are simply relocated. The long process of appeals and generous financial compensation that Americans enjoy is absent.

Readers will already be familiar with the heated battles between advocates of individual freedoms on one side, and advocates of more centralized planning and social policies on the other. Here we wish only to emphasize the fact that there are two different *kinds* of goods at stake. Democratic freedoms, such as free speech and religious freedom, are important social goods. But to make rapid, effective responses to crises is also a good, as is limiting the excesses of individuals and corporations when they act injustly by taking more than their share or by profiting from destroying ecosystems. Humanity has an urgent need to construct political systems where *both* sets of goods are manifested. Unfortunately, we seem to have difficulty even constructing conversations where the topic can be discussed.

As a Communist country, China is structured in a way that emphasizes the needs of the community over the wishes of the individual. Put another way, in "socialism with Chinese characteristics" the wishes of individuals are evaluated in the context of national and community priorities. Where Americans have prided themselves on their "rugged individualism," Chinese civilization back to the time of Confucius (c. 551–479 BCE) has focused on *individuals-in-community*. Over the centuries, in times of national crisis, the central government has been able to implement rapid, sometimes massive societal changes.

Of course, top-down change occurs in Western democracies as well. Here, too, there are excesses, as when large corporations are able to influence the American Congress to pass legislation favorable to their financial interests and damaging to the interests of middle- and lower-class citizens. But there are also opportunities for citizens to influence governing bodies in positive ways, at least at the city and state level. At a recent ecological civilization conference in Claremont, California, for example, a political leader from Los Angeles spoke about the power of citizens in bringing about top-down change, stating that "when 400 people show up to City Hall, we listen." In a city of 4 million people, that's impressive.

The idea of individuals-in-community is also built into the idea of American democracy, however cynical one may be about the ways that Americans do or don't live it out. The motto of the United States is *e pluribus unum*, "out of many, one." Western democracy is supposed to work whether politicians are individually altruistic, thinking only about the well-being of the people as a whole, or self-centered, prioritizing their own interests. When the voices of hundreds or thousands, or hundreds of thousands, of citizens come together and "out of many" a unified voice arises, they can have significant impact in bringing about top-down change. The Civil Rights Movement in 1960s, and the Supreme Court decision that helped legalize marriage for same-sex couples in 2015, are powerful examples.

How can you get involved in helping promote top-down change? Use your vote and voice. Work to get the vision and concerns of ecological civilization into mainstream public discourse. Let your political leaders know what you care about and mobilize your friends who share your interests. In a representational democracy, a passive citizen is merely complicit in supporting the status quo. Beyond voting, the citizen of a city or state can write letters, speak at hearings, join or organize political action groups, or participate in demonstrations. There are countries where citizens have less effect on top-down decision making, and other countries where they can have far greater effects. The more local or regional the activism, the greater the effects.

The top-down strategy for building an ecological civilization entails having policies, laws, and regulations that encourage citizens and corporations to operate in ways that promote long-term sustainability and overall well-being. In the widest sense, this includes international policies and commitments, like those offered through the United Nations. Think of the crucial roles played by the UN Sustainable Development Goals and the 1992 UN Framework Convention on Climate Change, which has continued to meet as a "Conference of the Parties" (COP), as in the COP21 meeting in 2015 that produced the Paris Agreement, the first global accord to combat climate change.

National efforts at top-down change lack the symbolic power of a global resolution. But nations can craft and enforce legislation, whereas the United Nations has virtually no authority to enforce policy over sovereign nations. When the German Chancellor Angela Merkel pledged in May 2019 to make Germany carbon neutral by 2050, her government immediately began work on drafting legislation across multiple sectors that would help to achieve that goal. In the United States, legislation based on the "Green New Deal" vision, were it to be passed by Congress, would have a tremendous top-down impact toward building an ecological civilization in the United States.

But there's the rub. In many countries, and in the United States in particular, the prospects for serious action at the national level are bleak. In other places, limited national resources, or a crippled national government, are already making it impossible for countries to respond adequately to increasingly frequent and extreme climate disasters. When nation states do not or cannot act, cities have to become the primary unit for responding to present needs and for building sustainable structures for the future. In places where even cities are unable or unwilling to respond, local communities become the final unit for resistance and support.

Bottom-Up: Strong Local Communities are the Cornerstone of Ecological Civilization

While some city leadership takes the form of top-down approaches—especially in large cities like New York, Los Angeles, Beijing, Seoul, etc.—other kinds of action at the city level more resembles the bottom-up model for change. The bottom-up strategy for building an ecological civilization entails having individual citizens, and local organizations, take on the responsibility of acting on behalf of the vision of long-term sustainability and overall well-being.

When the bottom-up approach begins with individual citizens, it's about personal choices: Choosing to ride your bike to work instead of driving a gas car; choosing to eat meat-free one day a week, or two or permanently; choosing to turn off your air conditioner and opening the windows; or choosing to buy local vegetables from your farmers market rather than supermarket tomatoes transported two thousand miles, or grapes shipped in from the global south. The list goes on. Of course, one person's choices won't produce a global paradigm shift to a new kind of civilization. So how do personal choices of this bottom-up variety help us begin building an ecological civilization? It's often a matter of strength in numbers. One person's choice to go vegan will have little effect on the global system of agriculture. Now imagine one million people in New York City who stop buying meat and dairy products. What happens to the pepperoni pizza industry in New York? Those who wished to stay in business would need to change their product offerings. If one-eighth of the population were to swear off cheese, it would significantly impact the production and distribution of dairy products in the region. Soon the number of domesticated cows being raised in the area would decrease, and a paradigm shift in regional farming of livestock would be underway.

The bottom-up approach refers to local structures (cities, bioregions) or hyper-local communities (neighbors, extended families, villages), where change operates in a blended way. Broader top-down policies may be at work, but their influence is generally more modest at the community level, where the bottom-up influence of individuals and groups plays the primary role. When individuals and small communities are the agents of change, one finds new ideas for reimagining and restructuring society, such as experiments at introducing rural and agricultural practices into urban centers in ways that promote health and sustainability. Innovations introduced through the Transition Town movement (www.TransitionUS.org) offer clear and powerful examples. The hybrid nature of change at this level is the result of (and also contributes to) strong local communities that are semi-autonomous but never isolated—communities in which individuals have some real control over their lives and futures but make choices in ways that serve the good of others in their communities.

A top-down approach without support from the bottom will be met with resistance and will eventually collapse. Similarly, a bottom-up approach that isn't supported by policies at the top will be crippled and ineffective. A successful framework for building an ecological civilization requires both top-down and bottom-up change; it connects the grassroots with the grasstops, as it were. In some places, like China, the top-down approach tends to have the greater impact; in places like rural Idaho, a bottom-up model will likely be more effective. Ultimately, achieving structural change requires getting the tops and the bottoms—policy makers and local organizers—to work together as allies to bring about a more sustainable and equitable way of living on this planet.

Conclusions

This and the following chapter begin to lay out a roadmap toward civilizational change and identify some of the steps that people are beginning to take. Our goal has not been to construct the entire roadmap (see Question Eight), but rather to draw attention to the need for a sustained program of research and action. New scholarship and activism are expanding this program and deepening its foundations in thought and action. It is increasingly important to attain clarity on what the transition between civilizations means, what it entails and excludes, and what actions we can take now so that humanity—and the planet whose future this generation holds in its hands—will succeed at the transition. What makes this task so urgent is that humanity is already in this midst of this transition. And yes, as crazy as it sounds, as a species we don't yet have a clear sense of what the goal is that we are supposed to be shooting for or what roadmaps we should be using in order to steer in that direction.

Even in this brief summary, though, one begins to recognize how all-encompassing are the implications of humanity's movement toward an ecological civilization. As we saw in Question One, civilizational change has been a regular part of the rhythm of human history; each time it occurs, it represents a fundamental shift in the way humans think about and live in the world. A transition between worldviews—the focus of Question Two—casts every part of human activity in a new light. The present and following chapters dig deeper into these changes, moving from worldview-level concerns to their concrete implications for life and action.

For this new movement to have its long-term effects, and for it to answer the objections from critics and defenders of the current world order, sophisticated analyses and concrete proposals will be necessary. Various groups—in South Africa, Asia, Europe, and North America—are already engaged in these activities. To be successful, the project must also include specifics about lifestyle changes, lest non-scholars conclude that civilizational change does not affect them. If communities and organizations are to use ecological civilization as their overarching goal and organizing principle, they will need to be convinced that working toward this target will actually make a difference. The goals of the ecological civilization movement also need to be clear and compelling enough that they can be communicated effectively to people in vastly different cultures and situations.

Presenting the depth of the ecological crisis itself continues to be important, since it brings home to communities, societies, and nations the urgency of rapidly changing the ways we live with each other and with the planet. Expertise in the various sectors of society is also indispensable if we are to show how each one will be affected. But it's no longer enough merely to talk about the ecological crisis and the devastating consequences it will bring. People need hope; without it they will cease to push for change and will return to "life as normal." Deep hope springs up as one works for the long-term future of humanity, even beyond the possible collapse of the present system. Hope grows from helping to bring about changes *now* that contribute to the transformation of society. The ecological civilization movement must be based on sound scholarship, but it must also transition to hope-giving at these two levels.

The long-term goals as we have formulated them here express a version of the Seventh Generation Principle. That principle is based on an ancient Iroquois philosophy that, "In our every deliberation, we must consider the impact of our decisions on the next seven generations." The Constitution of the Iroquois Nation (The Great Binding Law) explains "seventh generation" philosophy as follows: "The thickness of your skin shall be seven spans — which is to say that you shall be proof against anger, offensive actions and criticism."³ We affirm the truth of this ancient principle: "the decisions we make today should result in a sustainable world seven generations into the future."⁴

Now that we have worked through the theories and practices of ecological civilization, one may wonder even more concretely what this new form of society will look like. We thus turn next to exploring examples of groups already putting the theory of ecological civilization into practice in the areas of community, economics, education, and others.

Endnotes

1. We are grateful to David Korten for this articulation.
2. Pope Francis, *Laudato Si'*, §139.
3. "7th Generation Principle," Seventh Generation International Foundation, http://7genfoundation.org/7th-generation/.
4. "What is the Seventh Generation Principle?" Indigenous Corporate Training, Inc., https://www.ictinc.ca/blog/seventh-generation-principle.

QUESTION 7

WHAT DOES ECOLOGICAL CIVILIZATION
LOOK LIKE IN PRACTICE?

We want to know what human society will actually look like as truly ecological practices transform an unsustainable society into a sustainable one. This chapter is devoted to that question. We explore some crucial examples of new practices across the various sectors of society, offering concrete examples of ecological civilization coming down to earth—how it gets a street address, as it were.[1] In each case, we connect the new ways of thinking with the concrete ways that these ideas are being put into practice by individuals and groups. We focus here on practices that individuals and small groups can adopt, even without the help of larger organizations.

Putting the Idea to Work

It should be obvious why this task is crucial. In responding to the previous six questions, we have repeatedly considered the connection between theory and practice. At this point, many want the discussion to become as concrete as possible: how is ecological civilization motivating new practices? What are organizations that are embodying the kinds of values that have dominated our discussion? What kinds of goals and practices do they engage in, and how are they doing it?

And which of these values can we begin to live out now—as individuals, religious practitioners, members of communities small or large?

We imagine a skeptic who has been reading this book. He or she is concerned: "Ecological civilization is a rather abstract term; can it really do anything? That is, is it concrete and specific enough that it can really give rise to, and guide, actions? Also, is it too idealistic? After all, if it's too utopian, it can never become a reality, and especially can't become a reality in today's world."

These are the right questions. Ecological civilization is now moving beyond abstractions and becoming an action word. Its various features bleed into one another, like two colors on a handmade shirt, creating an integrated vision for action in the present. Proposed changes can now be evaluated in terms of their implications for rethinking and restructuring every sector of society: economics and production, energy and transportation policy, the revivification of rural life, the construction of sustainable urban centers, education, community, culture and tradition, spirituality and religion. Real transformations in each of these areas are the foundations for a broader civilizational transition.

Economics and Production

The strongest bastion of individualism around the world today is economics—or, rather, a particular understanding of what human economic activity is. Humans have always produced and exchanged goods, though often not in equitable or just ways. In the modern period, some European theorists argued that economics—systems of production and exchange—will only work effectively if each consumer makes decisions thinking primarily of what will be in his or her own best interest. The sum total of all that we make and exchange, they said, is "the market"; and if we let the market function without interference, each person's work will be rewarded fully and fairly.

It is admittedly counterintuitive to think that the more individuals maximize their own personal gain, the better off others will be. This idea became so important to some modern economists, however, that they concluded that humans just are *homo economicus,* "economic man." We are self-interested producers and consumers first, and everything else can be defined and valued from that standpoint. What we now know as *consumerism* is a manifestation of this view that the economic individual, the free-standing consumer, is the fundamental unit of society. Probably no issue more sharply defines the difference between contemporary society and an ecological civilization.

Neoliberal economics notwithstanding, it seems obvious that humans, if we are to survive and thrive on this planet over the very long term, will have to learn to define values differently than by the measure of what an individual can acquire and consume. Is it not inevitable that the communities on which people depend will be undercut when they place their interests above the interests of the groups to which they belong? Of course, decentering the individual does not mean that the solution lies at the opposite extreme, collectivism, which focuses on the group at the expense of the individual. Sadly, many of the 20th-century debates about economics were repeated wrestling matches between opponents on the two extremes. In practical decision-making one has to look instead for dialectical solutions—in this case, in the working together of strong individuals within strong communities.

Humanity can't even begin to think of moving toward an ecological civilization so long as we continue to affirm acquisition as the core human value. The standard model of the "rational economic actor" presupposes that each economic actor will seek to acquire as much as possible for as little labor as possible and at the lowest price possible. Game theory in economics, for example, which is now being used to model an increasing number of social systems, is based on these assumptions. Fortunately, a newer school in economic theory, ecological economics, is now using communal and environmental assumptions to model sustainable systems of production and exchange.[2] As promising as these alternative models are, they face immense opposition from the currently dominant economic system. Banks and other wealthy financial institutions will lose financial power in the transition to alternative monetary systems. It is unlikely that they will voluntarily loosen their control over industry and government. And yet communities can only become stable if the profits from local economic activity remain within the region, rather than being transferred away as profit to transnational stockholder-owned corporations.

The success of local economies depends finally on the mutual strengthening of local communities and not, for example, on the amount of money that outside investors are willing to pay to acquire its assets. Here frugality, sustainability, and reciprocity are far greater indicators of local economic health. Net growth of asset value, for example, does not solve the problems of unemployment and wages that are insufficient for local families to meet basic needs. Sustainable communities are built upon "micro-economies" in which the local economic units, such as families and small businesses, and not large outside financial entities, are the major players. Quality of life in communities is inseparable from having productive and meaningful work that sustains the members of the local work force and their financial dependents. People naturally seek a secure place in a healthy human community in a healthy ecological context. Local communities are far more likely to accept the responsibility to find employment for those who can work and for meeting the essential needs of those who cannot sustain themselves in this way.

We have inherited a global economic system in which the most valued quest is still economic growth, which is commonly measured by quantity of market activity or the Gross Domestic Product (**GDP**) of a nation. In fact, for many, "value" is now synonymous with how much money you have. The more valuable something is, the more money can be exchanged for it. This is not only true for individuals but has been expanded as the means to assess the value of entire nations. As strange as it may sound, **GDP** has become the primary criterion for assessing a country's well-being. Yet this *quantitative* measure of well-being is incapable of *qualitatively* evaluating the well-being of citizens. We often hear it said that "time is money." This suggests a narrow perspective on the value of our time. In truth, "time is life."[3] When we begin to prioritize money over life, wealth over health, our value system is skewed. Money can only be measured quantitatively—more and less. By contrast, a full life is built on rich qualitative values, subtle distinctions between better and worse. While **GDP** can measure market activity in quantitative terms, it fails to account for qualitative experiences. You can't measure love or happiness with a calculator! Yet even happiness tends to be equated with wealth, defined as a higher standard of living and a surfeit of goods.

As humanity hits the planetary limits on growth and the role of healthy ecosystems becomes unmistakable, the focus will have to shift toward sustaining regional economic communities that live in a "steady state" of interactions with their local environment. Nations can play a crucial role in this ecological transformation by turning the policy focus away from GDP-based measures of growth and enacting robust community-based economic and social policies in their place. In the end, sustainable human societies can arise only when growth is redefined as progress toward a systemically ecological and equitable society.

It is not easy to ascertain how well we are succeeding at this goal, since economic measurements of happiness and well-being are still in their infancy. Nevertheless, great work on this front has already begun. A number of scholars and leaders are reframing ways in which societal flourishing can be measured. The Kingdom of Bhutan, for example, has challenged the prominence of the GDP with an alternative called Gross National Happiness (GNH). The GNH index was added to the Bhutan constitution in 2008. This index was implemented to gauge the happiness and well-being of the population.[4] Rather than stressing the importance of an abstract collection of monetary transactions, GNH strives to evaluate the effects of governance through the collective well-being of the Bhutanese people.

Gross National Happiness is measured by evaluating nine factors related to a person's overall happiness, including Good Governance, Ecological Diversity and Resilience, Community Vitality, Cultural Diversity and Resilience. A comprehensive 90-minute survey is conducted by the Bhutan government with individual households, using multiple layers of variables in order to assess its GNH. This alternative to GDP involves criteria that include both quantitative and qualitative variables, thereby creating a holistic approach to measuring progress.

His Majesty Jigme Singye Wangchuck, the former King of Bhutan, reportedly stated that "We do not believe in Gross National Product because Gross National Happiness is more important."[5] This is not to say that the GDP is unimportant or trivialized within the Kingdom of Bhutan, but it does show a different awareness than the perspective offered by modern global economies. While the well-being of citizens includes economic well-being, it can't be reduced to economic well-being. Whereas GDP is reductionistic, GNH is holistic. Rather than defining progress narrowly through the fluctuations in market activity, the Kingdom of Bhutan stresses both the economy and emotional wellness of its citizens. When progress is measured by financial gain alone, we ignore the most important parts of living—the experiences of feeling satisfied, being happy, and enjoying enriching relationships. "Try not to become a person of success but a person of value,"[6] as Albert Einstein once said. When nations measure success in terms of "sustainable human happiness" rather than increasing market activity, this is indicative of structural changes being made that can bring about civilizational change.

New forms of ecological economics, grounded in an environmental ethics and an organic understanding of human society, are already available to undergird these new measures of development, narrowing the gap between use value and exchange value. The intense resistance to ecological civilization stems, in part, from the addiction to economic growth and development. In the words of Peking University professor Huan Qingzhi, "The crux of China's environmental problem is the one-dimensional economic ideology of modernization development."[7] This ideology has increasingly become the political and social servant of "the logic of economic growth" and even of "the logic of capital" itself. The economist Herman Daly has called the worship of economic growth "growthmania," especially in cases where economic growth is regarded as "both the panacea and the *summum bonum*," the highest good.[8] Growth is seen by many as able to solve almost all important social problems, including poverty, unemployment, and crime.

For constructive postmodern thinkers like John Cobb, working in the Whiteheadian tradition, the current mainstream development model is both unsustainable and deeply problematic. In his book, *The Earthist Challenge to Economism,* Cobb listed some of the consequences:

> It has uprooted millions of people, separated hundreds of millions from their traditional communities, increased crime,

addiction, and family breakdown, slaughtered many of the poor in low-intensity conflicts, decimated the world's forest cover, depleted its fisheries, eroded much of its topsoil, and sped up loss of biodiversity. It is changing the world's climate in ways that are likely to cause critical problems for our grandchildren.[9]

Standard economic theory holds that the environment is part of the economy and needs to be subsumed under it so that growth opportunities will not be missed. Nature has instrumental value as the fuel that drives human production and economic growth. We think standard economic theory is wrong. In fact, the reverse is true: the economy is the part, and nature is the whole. Human production urgently needs to be integrated into the finite, entropic ecosphere so that the limits on growth will be visible. Similarly, traditional economists have distinguished *growth* (quantitative increase in size by accretion or assimilation of matter), from *development* (qualitative improvement in design, technology, or ethical priorities). By contrast, ecological economists advocate *development without growth*—qualitative improvement without quantitative increase in resource throughput beyond an ecologically sustainable scale. Our vision and policies should be based on an integrated view of the economy as a subsystem of the finite and non-growing ecosphere.

The ultimate end of human life on this limited planet, whatever it may be, cannot be unlimited growth. A better starting point for reasoning together is John Ruskin's famous principle that "there is no wealth but life."[10] We cannot even begin to think of moving toward an ecological civilization so long as we suppose that the human goal is essentially acquisitive. The ecological civilization movement begins by redirecting the economy toward a common good. We seek a sustainable future for humans who are living responsibly and diversely in our common home, *oikos*, the Greek root for both economy and ecology.

For an ecological civilization to be even possible, we must recognize that current economic analyses of society are inherently abstract. Societies, social contracts, laws, money—these are all abstractions, albeit ones that deeply impact our daily lives. They are examples of the fallacy of misplaced concreteness (see Question Two). This reification of money is one of the most problematic for building an ecological civilization. In their award-winning book, *For the Common Good: Redirecting the Economy Toward Community, the Environment, and a Sustainable Future*, Herman Daly and John Cobb demonstrate how many of the problems with our current economic theory and practices are rooted in mistaking an abstract economy for something concrete.

In sum, change must begin with a critical reevaluation of mainstream contemporary economics, including a challenge to several of its most central assumptions: that human nature is fundamentally "acquisitive," that growth models are the only viable form of economics, and that ecosystems are there to serve (short-term) economic success. The framework for long-term economic viability will be established instead by identifying what is necessary to retain healthy local systems. To make this change is to reorient economic theory and practice around the values of living communities in interaction.

An ecological civilization questions the relations between economic gains, progress, and happiness, turning our attention first to the dynamism of experiences and lived values. Using something like the GDP model to measure the success of friendships, one might be inclined to see how many "friends" someone has on Facebook. (This practice is more widespread than you might think.) But that number says nothing of the depth and quality of those relationships—their role in good times and in bad—nor about the overall impact they have on a person's identity and well-being. In human flourishing as it has been understood in cultural and spiritual traditions around the world, wealth appears as only one variable within the complex equation of life. Human flourishing invites us to look beyond narratives that are fixated on monetary measurements and to find new modes of meaning and importance in concrete relationships with family and friends, community, and nature.

At this point, you might be thinking, "So what? I don't use GDP to measure my family's well-being. I'm not a professional economist working on national indicators. What does Bhutan have to do with my day-to-day living?" While GDP and GNH as systems for assessing value and success may seem more abstract than practical, it's important to remember that we live each day in community. Communities are built on systems. The most dominant system in the world is the economic system of capitalism, for which **GDP** is the preferred measure. So what can you do; how might the insights of Gross National Happiness be applied on an individual level? Next time you go to the store, or browse on Amazon, try considering your purchase decisions in light of your overall happiness and the well-being of those around you. Next time you invest in stock, consider more than just the return or the quarterly gains, but the social and environmental impact of the companies you invest in. A great example of this shift in priorities is the Intentional Endowments Network,[11] which works primarily with colleges and universities on divesting from fossil fuel-related stocks and reinvesting in companies that prioritize sustainability.

Banking and Finance

Related to economics and production is the topic of banking and finance. It should come as no surprise that our current economic system is controlled by a handful of privately owned multinational banks. The primary objective of these private financial institutions is to maximize profits and protect their own economic interests. In connection with capitalism, our current private banking system has contributed to radical economic inequality—an unprecedented concentration of wealth in the hands of just a few. According to a 2017 report by Oxfam, the eight richest men in the world control the same amount of wealth as the poorest half of humanity (3.8 billion people). This inequality is not simply about money, but has implications for power and influence over politics, laws, societies, and even the likelihood of war. With private banks comes private interest. Often, those private interests are pursued at the expense of public interests and the well-being of the many. This is a systemic problem.

So, what are we do to? We are told that relying on private banks, many of which are "too big to fail," is necessary for a stable economy. The ecological civilization movement challenges this assumption. Imagine what the banking system would look like if supporting the common good were its foundational concern. What would it look like to have financial institutions that empower small business owners, homeowners, and local governments—banks whose mission is to provide for the long-term good of these groups rather than to bring profit to the bank's shareholders?

Viable models are already available. For example, "Public Banking" has been an important step forward in building smaller, localized economies that support local communities. Among the leaders in this movement are Ellen Brown and the Public Banking Institute (**PBI**).[12] After the economic crash of 2008, the need for alternative banking and monetary systems became crystal clear. In response, Ellen Brown founded the Public Banking Institute to promote a "banking and monetary system that supports sustainable prosperity for *all of us*."[13] As Brown often emphasizes, the leading success story in the US context is the first US-based public bank, the Bank of North Dakota. As a state-owned financial institution, the Bank of North Dakota seeks to "deliver quality, sound financial services that promote agriculture, commerce and industry in North Dakota." This bank is a financial institution with the primary goal of promoting regional well-being, not increasing financial gains for its private owners or stockholders. The Bank of North Dakota is just one successful example of public banks at the state level. The public banking movement involves a way to blow the whistle on the reckless practices of private banks (as was visible in the 2008 economic crash), while promoting alternative systems that invest in the health and well-being of local communities.

Chances are, you don't live in North Dakota. So how can you get involved with alternative banking and finance systems that promote the well-being of local communities? Consider participating in a credit union in your area. Better yet, see if you can help to develop an alternative local currency. A number of such systems of local trade exist in North America and England, modeled on the ancient systems of barter and trade that were widespread in earlier eras. The BerkShares currency used in the Berkshire region of Massachusetts offers an excellent example.[14] Perhaps you can find some local peer-to-peer cooperatives that promote a "commoning" system like those promoted by Gar Alperovitz and the Next Systems Project.[15] As the fundamental flaws in our current banking and finance systems become more visible, new creative alternatives are beginning to emerge.

Education

Education is an integral part of life. In the broadest sense, education encompasses every form of human learning. Some education is formal (think schools), while other learning takes place informally through everyday experience. Rethinking education may be the single most important practical step in the movement toward a society based on ecological principles.

Four changes represent starting points for an ecological model of education: discarding the myth of value-free education, challenging the current stress on narrow specialization, focusing on the exploding ecological crisis, and moving questions of sustainability to the center of the curriculum. Let's consider them one at a time.

Today, most formal education is oriented toward "book knowledge" and career development. The rise of so-called "value-free" research claims to study and present the facts of the world impartially, free from the stain of subjective values. Such modes of education dominate the modern world. Implied in value-free education is the view that the world is constituted by facts separate from values. This fact-value dualism has contributed to a bifurcation of nature into objects (facts) and subjective values. But when we understand the world as a community of subjects rather than a collection of objects, we begin to see how value is inherent in all facts.

For example, global warming is a fact. To treat this fact as value-free (neither good nor bad) implies that one is indifferent to the extinction of species and the quality of lives of future generations. *Value-free education is not free of values, it is simply free of reflection on values.* Such education is incapable of responding to the most critical problems of our world, since to recognize a fact like global warming as a problem is already to make a value judgement. By contrast, we need models of education that prepare students to find creative solutions to global problems, starting in their own rural or urban context—in their own backyard, as it were. Since the problems are comprehensive, the solutions will need to be collaborative. Therefore, we need models of education that train people to work with others on complex problems. We need models of education that prepare people for success in a society based on ecological principles, which means learning how to live in harmony with nature and one another. Without pedagogical methods that promote reflection on values, students cannot be prepared for these new tasks.

Solving real-world problems is different from advancing research in a highly specialized field. But challenging the current structure of specialized disciplines is not just a matter of changing some educational policies; it requires us to formulate a vision for new ways of organizing knowledge and teaching in general. Because the commitment to ecological civilization is deeply value-laden, any claim to value-free education will necessarily blind students and faculty to the urgency of action in the face of the crisis.[16]

So long as scholars stay in their own disciplines, responses to the climate crisis will be fragmented and progress will be marginal. A powerful and comprehensive solution can only emerge out of integrative cooperation across all fields of specialization. It's not as if expertise is the problem. Not everyone can be an engineer; not everyone can be a physicist; not everyone can be a surgeon, a musician, or an educator. We need each of these experts—and many more—if we are to build a sustainable, ecological future.

The values of wisdom and appreciation must be placed above (or at least integrated into) specialization within a single field. Classically, education focused on whole persons, their broader formation, and the overarching goal of developing their humanity (*humanum*). Interconnectedness preceded analysis and specialization. The mission and structure of contemporary universities cannot serve this task without radical revisions. The economic model of higher education requires a large number of more or less independent schools, each forced to seek its own funding in order to survive, often from businesses—a process generally more driven by market forces than by societal values.

The reforms cannot first begin at the college level. Schools that are located in rural settings and that encourage students to think and explore in open-ended ways are usually far more successful than large urban universities at producing whole persons who can formulate their own core values. K-12 education can include growing gardens, learning about (and experiencing!) ecosystems, and studying forms of responsible living. High schools and colleges can center on highly interdisciplinary or transdisciplinary programs, include appropriate internships, and use the notion of ecological civilization as a framing concept. Graduate programs can begin with actual values issues in the world and provide advanced skills for addressing them.

Given that reforming society in an ecological direction is necessary for our survival, educating future citizens so that they have the required skills becomes the highest priority. Taking even the first steps is impossible, however, if we don't first reintroduce the category of value as a (or the) crucial component in the educational process.

The comforts of the modern world are not readily available to everyone, but its effects are felt everywhere. The varied environmental and social contexts in which we find ourselves invite us to recognize that different contexts require different solutions. In each region the interactions of social stratification and exclusion, racial and ethnic differences, cultural belief and practice, and economic systems are unique. In each individual case they give rise to a unique set of issues, which is why they require distinctive, localized solutions, especially in rural communities. Instead of a one-size-fits-all model for teaching ecological principles, education for long-term sustainability takes highly specific forms. In healthy rural settings, for instance, patterns of life that are in harmony with nature are easier to find and thus to teach. Robust rural communities offer children direct examples of lived ecology, whereas teachers have to work hard to find similar examples in poor inner-city settings that are devoid of parks and gardens.

Rural Education and Food

People attempting to break away from lifestyles dominated by capital-driven consumption and those attempting to make rural communities more efficient know the difficulties of this task. A particularly helpful model of success comes from the Asian Rural Institute (**ARI**), based in the Tochigi prefecture of Japan. This educational non-profit organization helps rural and grassroots leaders learn the ins-and-outs of effective sustainable agricultural practices and communal living. The participants do not earn an academic degree or attend classes in the traditional sense. Instead, the program cultivates the next generation of local leaders through hands-on learning and educational experiences, impacting the wider world through the lived value of communal living. ARI participants come from around the world to take part in this experiential educational program.

ARI's nine-month training program is built on the three pillars necessary to support a thriving community: servant leadership, "food-life," and community building.[17] Each one of these pillars points to a paradigmatic shift in perspective. Servant leadership combats the competitive nature of modern thought and stresses the mutual flourishing of various segments of society. Foodlife reestablishes the centrality of food as an indispensable fact of life, highlighting the reciprocal relationship between food and life itself. Finally, community building highlights the strengths of communal togetherness. The ethnic and cultural differences of the participants become resources in working toward the shared vision of sustainable community building. After each transformative program comes to end, the men and women return to their respective communities, seeking to act as catalysts for positive change.

ARI teaches practical, day-to-day knowledge that can be applied around the world, tying theory and practice together in intricate ways in order to create flourishing communities. Its pedagogy counters modern educational systems that overemphasize individuality, competitiveness, and profit-driven standards, which will not transform civilization for the future but only perpetuate the current unsustainable trajectory. One of the primary ways to educate adults and children about ecological living is to use curriculum and pedagogical methods that reflect on values and promote embodied knowledge for mutual flourishing.

These types of educational models represent a shift away from siloed specializations to integrated practical wisdom essential for more just and sustainable ways of growing and sharing food. An example of this model being applied in some 160 countries is visible in the work of Slow Food.[18] The direct goal of the organization is to support good, clean, and fair food. It works "to prevent the disappearance of local food cultures and traditions, counteract the rise of fast life, and combat people's dwindling interest in the food they eat." As an educational approach, Slow Food helps to teach people in each region where their food comes from and how their food choices affect them, their region, and the broader world. At the same time, it emphasizes systemic factors. Its educational goal is to show how "food is tied to many other aspects of life, including culture, politics, agriculture and the environment. Through our food choices we can collectively influence how food is cultivated, produced and distributed, and change the world as a result."[19]

Educating for life, rather than specialized research, is a mark of education for ecological civilization. In particular, teaching about the production, distribution, preparation, and eating of food is fundamental to a different way of living on this planet. After all, everyone eats. Although we've heard our children say "I'd just die without Netflix," the truth is that there are very few things that we'd actually die without. Among the ones that top the list are breathable air, shelter from the elements, water, and food. When these things are in ready supply, we tend to give them very little thought. Despite the fact that food is one of the most important conditions for human survival, it is given only slight attention in most of the K-12 curriculum. This has resulted in generations of people who are unprepared to make informed food decisions on a daily basis—even where these decisions radically impact their health and quality of life.

Today fewer than 2 percent of Americans live on farms,[20] which contributes to a growing disconnect between the production and consumption of food. Nearly half of Americans say they never or rarely seek information about where their food was grown or how it was produced, and don't typically know whether the food they consume is genetically modified—even though **GMO** products are currently found in over 75 percent of packaged food in the United States.[21] Additional evidence shows just how uninformed Americans are regarding food. One study showed that 7% of all American adults (17.3 million people) believe that chocolate milk comes from brown cows.[22] Another study commissioned in the early 90s by the US Department of Agriculture found that nearly 20% of adults did not know that hamburgers are made from beef.[23] That's not even to mention recent studies that demonstrate how uninformed American children are, such as not knowing that pickles are cucumbers, that onions and lettuce are plants, that French fries come from potatoes, and that potatoes grow in the ground.

How much do you know about the food that you eat? On the model of Slow Food and **ARI,** you can learn more about the food in your life by being in conversation with local farmers. Consider purchasing your fruits and vegetables from a local farmer's market. There are often groups nearby—Amy's Farm[24] offers an excellent example—where you can volunteer to help with the growing and harvesting. Or maybe you can look for farm-to-table restaurants in your area. Perhaps you can connect with a local gardening center to see if there are any classes or groups you can join. As a start, you can watch YouTube videos or join a Facebook Group where you can learn the techniques of composting and "urban homesteading," which means growing fruits and vegetables, and even raising chickens, in the yards of even small urban homes.

Even a small start can make a difference. Learn about the food cultures and traditions of your family and ethnic history. Consider getting involved in a local religious group that is involved in growing and distributing food. After all, many spiritually based groups pay close attention to the values that are the basis for human action, which makes them a particularly good place for informal value-centered learning. And, whatever other steps you take, make sure you are purchasing locally grown fruits and vegetables rather than those that have been shipped in from South Africa, South America, and Asia.

Religion and Spirituality

Modern societies tend to erroneously view religion and spirituality as archaic modes of thought, superstitions that are concerned with matters that have no bearing in the actual world. They are often deemed a distraction, in Marx's famous phrase, an "opiate of the masses." Yet, religion has always been, and remains, an important part of human society and civilization. If we are to have a new type of civilization committed to the common good, we need to identify positive roles for religion and spirituality.

Today, more and more people are realizing the crucial role that tradition and religion will play in creating an ecological civilization. In the words of Pan Yue, "From the Taoist view of Tao respecting nature, to the Confucian idea of humans and nature becoming one, to the Buddhist belief that all living things are equal, Chinese religion has helped our culture to survive for thousands of years. Chinese religion can be a powerful weapon in preventing an environmental crisis and building a peaceful harmonious society."[25] Christian theology, castigated by Lynn White in 1967 for being inherently anti-environmental, has now inspired a vast array of books on eco-theology, sustainability, environmental ethics, and related topics.

Can the world's religions help contribute to ecological civilization? The answer is fairly straightforward: they can so as long as they continue to remind us that we live together within a larger community of life on Earth—the Earth community—which is our common home. In fact, without support from religious and spiritual communities, there is no chance of building an ecological civilization. More than 85% of the world's population identify as religious. That's more than 6 billion people, making religion the largest actionable constituency on the planet.[26] If you want to bring about significant change on a global scale, you'll need to work with religious communities.

The global perspective is crucial. When religions are provincial and competitive, they spawn division and ultimately violence. When they stress community over chauvinism, they are a uniquely powerful voice. Religious and spiritual traditions provide positive guidance as long as they inspire people to live with respect and care for the whole community of life—including those at the bottom of human society. They have the capacity to inspire people to build local communities that are creative, compassionate, participatory, ecologically wise, and spiritually satisfying, with no one left behind. And they can inspire leaders to develop public policies that are conducive to these ends.

In order to succeed in their role, religious leaders must also be able to formulate their reasons for action on behalf of the earth and its most endangered inhabitants—something similar to what Pope Francis achieved in his ecological encyclical *Laudato Si'*. As value-oriented communities, religions are well-positioned to challenge key assumptions of contemporary life that prevent humans from appreciating their common home, building sustainable communities, and living with respect for all living things. With their worldview-level beliefs, supporting practices, and ability to transform individuals and groups, religious and spiritual traditions are indispensable for helping humanity complete the transitions that now lie before us: to move

- from a shallow empiricism that focuses only on sense experience to a deep empiricism, so that we may appreciate the wide range of experiences that yield wisdom concerning the natural world;
- from an overemphasis on individualism to a recognition that we are persons-in-community, whose well-being depends on the well-being of greater wholes;
- from an anthropocentrism that focuses primarily on the value of human life to a recognition of the value—and indeed a love of—all life (*biophilia*);
- from exclusivist loyalties to our nations and ethnic groups to world loyalty; and
- from selective compassion to inclusive compassion.

It is impossible to overestimate the transformative role that the world's religions and indigenous lifeways can play if they take the lead in living out this commitment. They make the call to individual transformation and to corporate action, without focusing only on individual choices. The wisdom needed in the transition between civilizations that is already underway must be an embodied wisdom—a felt sense of the world that is ritualized and translatable into practical action.

One often encounters the assumption that Christianity will play the leading role in the transition that lies ahead, perhaps assisted by

Judaism and Islam as the other members of the family of Abraham. For many reasons, this is unlikely. It is crucial to be thinking of the full range of religious traditions, each with its own particular strengths and ways of influencing both adherents and non-adherents. Consider the following examples.

Hinduism invites us to embrace plurality itself, recognizing what Diana Eck calls the "manyness" of God—that is, the multiple faces and names by which the divine reality, understood in personal or transpersonal terms, can be known and felt. Insofar as an ecological civilization includes a healthy respect for diversity, Hinduism offers its special voice.

Buddhism invites us to recognize that, when push comes to shove, it is not the amount of material goods we possess that gives us our freedom, but rather the twin virtues of wisdom and compassion that make us whole. Insofar as an ecological civilization is built upon a wise recognition of the interconnectedness of all things, and a compassionate response to all living beings, Buddhism offers its special voice.

Jainism encourages us to live with deep *ahimsa*, with radical non-violence, thereby widening our sense of who deserves our compassion. Perhaps no religious tradition embodies respect for life more than the monks and nuns of Jainism, who wear masks over their mouths to avoid harming microorganisms in the air they breathe. Insofar as an ecological civilization requires inclusive compassion, Jainism offers its special voice.

Taoism encourages us to sense a *qi*-like energy throughout the whole of the universe, incarnate, among other places, in our own bodies, to which we can awaken and live healthy lives. Insofar as an ecological civilization requires sensitivity to the energetic dynamics of life and sensitivity to the inner lure toward holistic health, Taoism offers its distinctive voice.

Confucianism invites us to see human beings within a larger cosmic context—a trinity of heaven, earth, and humanity—and then to live within our own context in humane, reciprocal ways, informed by a sense of *li* or appropriateness. Insofar as an ecological civilization entails this kind of relational living, this sense of being a person-in-community, Confucianism offers its unique voice.

These are but a few of the South and East Asian traditions. Other "religions of the book," as well as indigenous lifeways, need to be considered as well; and even in these five cases we have mentioned only a few of the gifts they offer. The contributions of the various traditions are complementary, although modern thought since the 16th century has turned them into sets of incompatible truth claims that need to be either condemned as false or forcefully subdued for the sake of the one true religion. In fact, though, it is as a collection of related paths to compassion that the religious traditions can offer the strongest hope to a world in transition.

In recent years, many faith-based organizations have been mobilizing for sustainability and eco-justice. Our current environmental situation has drawn an amazing response from across the religious spectrum. Consider, for example, GreenFaith, an interfaith coalition for the environment that encompasses a number of the world's religious traditions. This religious alliance develops and implements various environmental programs to disseminate the shared vision of religious commitments as a powerful conduit for environmental stewardship and ethical responsiveness. GreenFaith offers an array of educational resources that highlight the responses that several religious traditions have brought to the current state of the environment.

Since its inception in 1992, this New Jersey-based nonprofit has transformed persons and religious institutions through its three guiding principles of Spirit, Stewardship, and Justice. They regularly develop new programs that highlight humanity's spiritual connection to Earth and creation. This conviction is reflected in their educational programs, one of which (the GreenFaith Fellowship program) educates religious leaders about the importance of the environment and calls attention to the current global crises humanity is facing. The goal is to give religious leaders the tools to make their houses of worship sustainable, and to advocate for an ethical consideration of others and nature. Their work enhances religious life by making it an integral component of the world. Religious beliefs and practices become manifest in the call to protect the environment and be critical of human consumption habits.

Besides offering programs on behalf of environmental literacy for religious leaders, GreenFaith organizes religious communities around central ecological concerns, such as their "Divest & Reinvest Now!" campaign, which urges religious denominations and institutions to divest from fossil fuels and reinvest in clean energy. Another project is GreenFaith's environmental leadership program, a two-year certificate program, which guides religious congregations in sound environmental practices, many of which provide financial benefits to the congregation.

An ecological civilization embraces the spiritual qualities of life. The religio-spiritual dimension lends itself to an ecological vision that is able to imbue the universe with ethical imperatives and a sense of responsibility for creation. It encourages us to comprehend the universe differently—not as a lifeless deposit of resources designed to meet human needs, but as a fragile system with its own intentionality and inherent worth. Often, religious persons talk about denying the "self" in order for the true "Self" to come into focus. This shift aligns with an ecological civilization that seeks to expand our circumference of awareness and our own place within it. As a result of widening our understanding of what the religious perspective encompasses, the tasks of working for the betterment of the society and the environment become pivotal undertakings for religious individuals and institutions.

So, what does this mean for you? If you are part of a spiritual community, perhaps you can start a small group focused on environmental justice, or lead a gardening project on the land of your church, synagogue, temple, or mosque. Perhaps you can invite business leaders, scholars, farmers, and environmental activists to serve as guest speakers for your community. Or maybe you can serve as a representative at the next Parliament of the World's Religions and participate in the Climate Action or Justice sections.[27] If you aren't part of a spiritual community, consider spending some time outdoors to connect with nature on a deeper level, and then finding others who share this conviction and who are willing to work with you. Or, if you are part of a secular nonprofit working on environmental justice, consider partnering with local religious and value-based groups on shared projects.

Agriculture, Tradition, and Technology

When some people think of an ecological civilization, they envision a world prior to modernization. They assume people pursuing this new kind of society are anti-development, that they want society to return to a pre-industrialized age, as if the benefits of our modern world have gone unnoticed. This is simply not the case. The healthy human society of the future will embrace technological advances and the desire to use these developments for the common good. We need to utilize the benefits of modernity, while changing the guiding principles of society toward more holistic and sustainable ends. Ecological civilization is not about "going back" to some golden age, it is a way forward toward a future of planetary flourishing.

A leading organization demonstrating the fruitful interplay between technology and environmental sensitivities is The Land Institute in Salina, Kansas.[28] Founded in 1976, this revolutionary organization has sought to redefine agricultural practices by understanding and replicating the symbiotic relations that naturally occur within the ecosystem. Simply put, they are promoting models of agriculture that work with nature, rather than against it—mimicking the environment's natural patterns of mutual flourishing. In contrast to industrial agricultural practice that seeks short-term gain through the use of pesticides, annual monocultures (an agriculture practice that plants only one crop in a given area, which needs to be replanted annually), and synthetic forms of nitrogen, the Land Institute under the leadership of Wes Jackson has promoted the use of perennial polycultures with a long-term concern for the health of the soil.

By utilizing the principles of evolutionary and ecological science, as well as applied skills of agronomy (the science of soil management and crop production), the Land Institute has breed a new perennial grain (Kernza®) that does not disrupt the natural cycles of the environment. By itself, this new grain does not entail sustainability, since Kernza can be used as part of a monoculture system. Therefore, the scientific development needs to be coupled with a change in worldview that values the long-term health of the soil and embraces a polyculture agricultural practice capable of producing ample food, while eliminating the need

Question #7

for the harmful practices of industrial agriculture. The result is both healthier soil and food security. The Land Institute's reorientation of agricultural methods successfully demonstrates the power of science in the service of ecological civilization.

Technology, then, is not at odds with sustainability; and yet we cannot rely on technology to fix all the problems with our world. The Land Institute demonstrates specific new paths for the agricultural industry by using scientific advances, such as the development of new kinds of seeds, at the same time that it reconceives humanity's fundamental relationship to nature. The result is a more ethical and environmentally responsible way of farming, which promises to offer new prospects for healthy, organic food items that are produced in a responsible manner.

Let's assume you aren't a farmer, and you don't plan to plant several acres of this new perennial grain, Kernza. How can you act in ways that will support these new models of agriculture and appropriate uses of technology? Perhaps you will decide to convert your lawn into a small garden. Perhaps you can buy locally grown organic produce, supporting agriculture that is not based on petroleum-based fertilizers. Perhaps you can eliminate animal products from your diet—meat first, then poultry and fish, and perhaps dairy as well. If you do eat meat, perhaps you can purchase sustainably farmed free-range meat from chickens, whose environmental impact is far less than cows and pigs. You could also consider the use of other important developments in technology. This might include adding solar panels to your home, or trading in a car for an electric bike. Of course, technology alone is not enough. Ideally, you will find ways to reduce your overall energy consumption, such as retrofitting your home. (A number of organizations exist to assist with this project, such as the Community Home Energy Retrofit Project.)[29] Even opening a window rather than using air conditioning helps. Each day, we eat food and use technology. We must begin to reflect on our consumption choices.

That ecological civilization is future-oriented with an openness to technological advances does not mean that we should reject tradition and culture. Creating an ecological civilization in general, and implementing environmental laws in particular, requires support from the specific wisdom traditions of each culture. One cannot study a nation's culture apart from the stories it tells itself about its past and its aspirations for the future. This is equally true of families and other societal institutions. An ecological civilization urgently requires a range of specific narratives that locate us all in one human history, including stories that locate that history in its context of the natural world as a whole.

Many peoples have been encouraged to leave their traditional values behind, which has raised the danger that meaninglessness or nihilism will result. This happened during the Cultural Revolution in China and can be seen in a number of trends that have emerged since that time. Succeeding in the global economic race pressures Chinese citizens, as it does other people, to abandon the spiritual resources of their own traditions, such as respect for heaven (*tiān*), acting in light of the oneness of nature and humanity (the Tao or "way"), continuing the tradition of thrift, and practicing the wisdom of "cherishing all living things." Constructive postmodern thinkers who are appreciative of tradition see these forms of received wisdom and tradition as valuable treasures that can provide strong moral support for ecological civilization, as well as for environmental policies that lead to effective environmental laws and constraints.

A number of other insights about culture will also play a key role in developing and supporting the move toward an ecological civilization:

1. A requirement of all cultures is food. After all, a core element of agriculture is culture. What is eaten, and *how* it is produced, distributed, and prepared, tell us a great deal about the meanings that shape the culture. Decisions about the production and consumption of food are not only at the heart of sustainability; they also contribute to personal enjoyment and social well-being. Ecological consciousness bears heavily on this part of our lives.
2. No culture can claim to be ecological that does not reward and respect those whose labor enables it to flourish.
3. We have become accustomed to the deep conflicts between issues of color, class, and gender, but in an ecological

civilization such distinctions cannot be drawn so sharply, and certainly not as the basis for social hierarchies.

4. However strong a group's cultural identity, in today's world it needs to remain open to the plurality of cultures around it.
5. Cultures are not static; variety, complexity, and change are characteristics of all cultures that endure.
6. A constant in human life is that it ends. This realization of mortality shapes reflection on meaning. Traditional thought and practices affirm death as well as life, as does the study of ecological systems.
7. Ecological civilization requires drastic criticism of current activities but will always call for self-criticism as well.
8. A reintroduction of beauty as a public value will serve to break the exclusive control of economics as the determiner of worth. If a society is to move toward a paradigm that supports life and that is ecologically sound, it must commit to beauty as a central value and organizing principle.

In short, the knowledge and appreciation of one's culture will not diminish but increase in importance in the transitions that lie ahead. After all, cultural commonalities knit a community together within a particular historical tradition; they give it a sense of a common mission and support activities consistent with the mission. Shared narratives provide a shared identity, offering reliable guidance as challenges arise. It is probably impossible for a group of people to hold together as a community, or for individuals to accept sacrifices on behalf of the group, without a sufficiently strong sense of shared cultural identity and the values that are concomitant with that identity.

Community

One of the most promising shifts towards an ecological civilization is reconceiving the role of community. Today we live in a paradoxical world. On the one hand, the technological revolution has connected us to others and to the planet in ways never before imagined by humans.

Information is instantaneously shared throughout the world at a click of a button. Humanity has never been so connected, and yet we have never been more alone. The new technological connectedness is not benefiting the most fundamental social units, the family and community. Suicide rates continue to skyrocket, while those who live in isolation in crowded cities often don't even know the names of the people with whom they share an apartment wall. As humanity becomes more and more engrossed in the technological revolution, human relations are breaking down like never before. Something is amiss.

What has happened to the social fabric of human communities? Why have they deteriorated during the modern age? These are not easy questions, but some responses are noteworthy. Global modernization has effectively relegated traditional concerns and communities to the wayside in favor of a single homogenous culture based on an economy-driven pattern of globalization. So imagine that, instead of uncritically joining the dominant culture of modernity, different communities begin to use their distinctive identities as ways to demonstrate that there really are alternatives.

Taken as a concrete guide, this dynamic highlights the need to rebuild community life. When people understand themselves as members of a community—in a place, standing in relationship with the land community around them—they start to see themselves as individuals-in-community. The process involves restructuring our social lives in ways that strengthen our sense of belonging, which means that we strengthen our sense of commitment to those with whom we live and to the places where we live.

The best and perhaps the only way to achieve this goal is to more strongly emphasize both our local identities and our local involvements ("localism"). Yet, just as we should be careful not to privilege community at the expense of the individual (or vice versa), we should also avoid embracing localism to the exclusion of the global. We strive not only to construct strong, healthy local communities, but also to build outward from them, connecting local communities into regional communities; recognizing that each local community is part of a community of communities. Local communities will have relatively self-sufficient forms of trade, but there will also be economic issues that require cooperation with neighboring communities. Healthy local communities will have as part of their basic self-understanding a respect and appreciation for other communities and their citizens. Rooted in their own local and social context, each will learn to coexist ecologically with the others and with their watershed, coast, or mountain region.

An ecological civilization may include a few extreme individualists, but rugged individualism cannot sustain an interconnected civilization over the long term. The complex dynamics of interdependent groups are incompatible with a world in which each individual sees himself as the primary unit. For societies to survive, the demands of individuals must be tempered by the needs of the group as a whole. Practically, this requires our learning to experience the world as a collection of societies rather than as an aggregate of atomic individuals. The most healthy social order is one in which each social group considers itself to be a community that depends constantly on the thriving of its surrounding communities.

The first step, then, is to move beyond viewing society as simply a collection of individuals, reconceiving human society instead as *persons-in-community, living for the common good.* This deep truth is well expressed in the traditional proverb, "It takes a village to raise a child": all of us as persons are formed by the intimate human communities that birthed and raised us. Each of us becomes a full person only in the context of community, and societies become authentic communities only as the people who make them up become stronger contributors to the whole.

Once we break free of the myth that satisfying individual wishes is the foundation of society, we learn to judge ourselves and those around us from the standpoint of their contributions to their family, neighborhood, village, or region. The locus of value changes. If I am always already a member of communities and would not exist as a person without them, then it's an illusion to imagine myself sitting apart from them, like a king on a hill, asking each group how they are benefitting me. We put our efforts into the thriving of our communities because we only become full persons through them.

One example of people reimagining community is the nonprofit organization Local Futures.[30] Local Futures seeks to educate developing communities about alternative options to inspire healthy local communities. They stress developments "away from dependence on a global economy dominated by huge corporations and supranational institutions, and toward economic structures that are more decentralized and diversified, with a much smaller ecological footprint."[31]

Founded by Helena Norberg-Hodge, a recipient of the Right Livelihood Award (also known as the Alternative Nobel Prize), Local Futures' first project started over 40 years ago in Ladakh, also known as 'Little Tibet,' in the Western Himalayas. At that time the region was facing enormous economic pressure to modernize and adopt global customs and etiquette. The allure of Western modernization was appealing for many, but leaders knew that their cultural and environmental context would not be able to sustain it. In the hope of undercutting the global narrative, Helena started the Ladakh Project, now known as Local Futures, and worked with the inhabitants of the region to develop their own narrative to carry them into the future.

Decades later, the effects of this project continue to be felt. By helping the people of Ladakh and others to search out alternatives to Western development models, Local Futures has helped communities maintain their unique identities within the shared vision of sustainable development. Local Futures is an exemplary model of seizing an alternative, insofar as it recovers the centrality of local communities as the focal point of society. Communities are built on the strong foundation of mutually beneficial relationships for the common good, imbued with indigenous wisdom, and supporting a vibrant culture.

Let's assume you're not planning to move to the Himalayan mountains to join the Ladakh community. How can you work to strengthen your own local community? In truth, our world is comprised of communities of communities of communities (and so on). Work out the implications of viewing your own immediate household as a microcosm of community: it's a part of a neighborhood, which is a part of a city, which is a part of a region, state, and nation. Today, in an age of globalization, the entire planet has become a community or communities that are ultimately grounded in local communities. Leaders like Norberg-Hodge encourage thinking globally while acting locally.

That can involve a range of actions—from buying produce from local farmers or utilizing services from small local business owners, to arranging neighborhood potlucks or community service projects. We encourage you to get involved with groups like Habitat for Humanity,[32] or to volunteer for a community center or after-school program, or to coordinate the creation of a community garden. In the end, healthy communities are the foundations of an ecological civilization.

Conclusion

Each of the sectors and organizations that we have explored in these pages is a potential source of actions that promote resilient local communities. Groups like this are real-life experiments; they are responding to the greatest mandate for action of our day: to begin to construct the fully sustainable civilization that will emerge as the present one dissolves, as all things that are not sustainable must inevitably do. We have no choice but to use the world that we inhabit today, and the world as it has been across history, as our reference point.

The purpose of this chapter has been to make explicit the various ways that ecological civilization is already becoming a reality, and the concrete actions that each of us can take to support this transformation. As we have seen again and again in the previous chapters, ecological civilization is not simply another word for environmentalism, nor is it limited to sustainability as the word is usually used—although the new society must certainly be sustainable! The examples explored in this chapter demonstrate the kinds of specific things that we must do in order to move toward a sustainable and just way of living.

Transformative action is crucial. And yet, beyond what we can *do*, the vision of ecological civilization also calls us to consider who we can *become*. It's not merely about having more or doing more, but about *being* more. This distinction may seem subtle, but it makes an extraordinary difference. Seeking to be an "eco-person" speaks to the deeper motivations behind the kinds of actions we have been exploring. The growth of eco-persons is connected with the qualitative measure of meaningful relationships, and not simply with quantitative measures of carbon, money, or degrees in temperature. Who we are is *more than* what we do, even though it obviously includes what we do. It's this something more that lies at the heart of the ecological civilization movement.

The examples that we have explored turn our attention upward, forward, and downward:

- *Upward,* because they help us to focus on the long-term goals. Even policy experts and grassroots activists can fall to the temptation to concentrate so much on individual trees that they lose track of the forest that is our long-term goal.
- *Forward,* as we begin to introduce the kinds of changes into today's society that can move us toward a sustainable civilization. In each sector we were able to list the steps, one after the other, that will lead to this new kind of a civilization—the outlines of a new roadmap. The truth is that no one fully knows where we are heading. So this exercise is indeed like sketching a map of uncharted terrain for the first time. As we have worked through these seven questions, a rough map has begun to emerge. Year by year, as the ecological goals (necessities!) become more clear, more details can be added to the map. Still, only as we begin walking the path does the destination become clearer.
- *Downward:* Think of the famous adage in management circles: when you have 80% of what you believe you need to make a decision, that's when you have to start acting. Looking downward means acting now to bring the big ideas down to earth. Sometimes it's literally looking at the earth beneath your feet, which may need water, or a garden, or a cleanup from industrial pollution; or it may need to become a playground, or to be given to the Nature Conservancy so that it can be permanently designated as a national park.

What's hopeful about the idea of ecological civilization is that it does not require us first to imagine a utopia on earth and then try to try to build it in reality. It's much more a series of concrete tasks than a dream of a perfect future; *it's a guide to specific action*, both at the policy level and in the individual ways that people live. Ecological civilization shouldn't be seen as a final, fixed destination or goal; it's always at the same time about dynamic ways of living in community.

Endnotes

1. We owe the expression to Eugene Shirley, president of Pando Populus; see https://pandopopulus.com/.

2. For some discussions on ecological economics, see Peter Brown's *Right Relationship: Building a Whole Earth Economy* (2009); John Erickson's *Economists are Morons* (2015); and Herman Daly's *Beyond Growth: The Economics of Sustainable Development* (1997).

3. We are indebted to David Korten for this phrase.

4. "Vision & Mission," Gross National Happiness Commission, https://www.gnhc.gov.bt/en/?page_id=47.

5. Tashi Dorji, "The story of a king, a poor country and a rich idea," published June 15, 2012, https://earthjournalism.net//stories/6468.

6. May 2, 1955, issue of *Life* magazine.

7. Huan Qingzhi, "Terminating the Growth Without Boundary," *Green Leaf* 10 (2009): 114–21.

8. Herman Daly, *Steady-State Economics,* 2nd ed. (Washington, D.C.: Island Press, 1991), 183. Daly borrowed the term "growthmania" from the British economist E. J. Mishan.

9. John B. Cobb, Jr,. *The Earthist Challenge to Economism: A Theological Critique of the World Bank* (London: MacMillan Press, 1999), 42.

10. See John Ruskin, "Ad Valorem," Essay IV in *Unto this Last: Four Essays on the First Principles of Political Economy* (London: Smith, Elder, 1862): "There is no wealth but life. Life, including all its powers of love, of joy, and of admiration. That country is the richest which nourishes the greatest numbers of noble and happy human beings." Kindle edition available through Project Gutenberg at http://www.gutenberg.org/ebooks/36541.

11 "Intentional Endowments," Intentional Endowment Network, http://www.intentionalendowments.org/.

12 See Ellen Brown, *Banking on the People: Democratizing Money in the Digital Age* (Washington, DC: Democracy Collaborative, 2019).

13 "Why Public Banks," Public Banking Institute, 2019, http://www.publicbankinginstitute.org/.

14 "Local Currencies Program," Schumacher Center for a New Economics, https://centerforneweconomics.org/apply/local-currencies-program/#BerkShares.

15 David Bollier, "Commoning as a Transformative Social Paradigm," published April 28, 2016, https://thenextsystem.org/commoning-as-a-transformative-social-paradigm.

16 Alfred North Whitehead made important contributions to a systematic rethinking of educational philosophy. See for example his *The Aims of Education and Other Essays* (1967); see also Robert S. Brumbaugh, *Whitehead, Process Philosophy, and Education* (1993).

17 "Our Training," Asian Rural Institute, http://www.ari-edu.org/en/our-training/.

18 See Slow Food at https://www.slowfood.com/.

19 "About Us," Slow Food, https://www.slowfood.com/about-us/.

20 "Food & Farm Facts: Resources," American Farm Bureau Foundation for Agriculture, http://www.agfoundation.org/resources/food-and-farm-facts-2017.

21 "How Many Foods Are Genetically Engineered?," UC Biotech, last updated February 16, 2012, http://ucbiotech.org/answer.php?question=15.

22 Nancy Coleman, "Chocolate milk definitely doesn't come from brown cows—but some adults think otherwise," last updated June 16, 2017, https://www.cnn.com/2017/06/16/us/chocolate-milk-help-trnd/index.html.

23 "20% of Americans don't know hamburger is beef. Ask them about chocolate milk!," Eideard, published June 22, 2017, https://www.

google.com/amp/s/eideard.com/2017/06/22/20-of-americans-dont-know-hamburger-is-beef-ask-them-about-chocolate-milk/amp/.

24 See Amy's Farm at http://www.amysfarm.com/.

25 Pan Yue, "Marxist Notion of Religion Must Catch up with Time," *Huaxia Times*, December 15, 2001.

26 "The Changing Global Religious Landscape," Pew Research Center: Religion & Public Life, published April 5, 2017, http://www.pewforum.org/2017/04/05/the-changing-global-religious-landscape/

27 See the Parliament of the World's Religions at https://parliamentofreligions.org/.

28 See the Land Institute at https://landinstitute.org/.

29 See the Community Home Energy Retrofit Project at http://www.cherp.net/.

30 See https://www.localfutures.org/.

31 "Global to Local," International Society for Ecology and Culture, https://www.localfutures.org/programs/global-to-local/.

32 See Habitat for Humanity at https://www.habitat.org/.

Question 8

Why Does Ecological Civilization Bring Hope?

The preceding questions on the meaning of ecological civilization have shown its richness and relevance. They took us through the history of human civilizations, the new sciences of ecology, and the underlying causes of the ecological threat. We have dug deep into the conceptual foundations and have cast the net wide to find leaders in other areas who are allies even when they are using different terms. Concerned that ecological civilization might be a merely utopian ideal, we have been able to show what it looks like in the real world as people begin taking the first steps in rebuilding society on ecological foundations.

People around the world are telling us that working toward ecological civilization brings them hope. How can this be, when the loss of hope has become the leading theme of our time—from baby boomers to Millennials and the often cynical members of Gen Z?

Why Hopelessness May Be Our Greatest Enemy

Those of us who often speak to audiences about climate change know the danger of discouragement. As we project the data onto big screens, we can feel the heaviness of the people in the room. To them it seems like each day more frightening data appears in newspapers around the world (which is true). We try to preserve their hope when we come to the things that need to be done. Sometimes we feel like soccer coaches: "Come on, you can do it; you can win this one!" After all, we tell them, hopelessness is our greatest enemy. We know that if we make the task sound impossible, people will give up; they won't do the things that the crisis calls for. Instead of being outspoken proponents and powerful activists for the planet, they will be paralyzed by overwhelming despair.

But after the public sessions are over, when the keynote speakers go out for a beer or coffee to speak as friends, our voices are softer and the discussion is much more sober. What are the latest data? What do the climate models predict? What are governments and businesses doing and not doing? In these conversations among experts you can sometimes feel the fear gripping your stomach. Will humanity open its eyes to the obvious and take the needed actions? Are humans fundamentally selfish? What if the direst predictions come true; what if we don't make it after all? Is it too late to avoid an economic and environmental collapse?[1] Living with the scientific and social details day in and day out, we also struggle with the sense of hopelessness. Physician, heal thyself.

Let there be no doubt about the data. Nations continue to fall short of the goals they set for themselves in the 2016 Paris Accord. At the same time, scientists are providing more specific lists of what needs to be done if we are to avoid a warming of 3 degrees Celsius (4.8° F) or more in the 21st century. And to cap it off, studies continue to show with greater and greater precision what will be the global impacts as the average temperature of the planet continues its apparently unstoppable increase. For each rise in temperature in each region of the world, we now know what will be the effects on weather, ocean levels, agricultural production, and available water supplies.

The stakes are high. If we fail to respond, we are looking at the destruction of the capacity for the planet to support life as we know it, affecting all 1.5 million species.[2] Yet the problems are immensely complex; they stretch across disciplines and sectors of society—from science to philosophy, from politics to economics, from agriculture to education. And the scale of the required reforms is massive. We're talking about fundamental changes to the habits of nations, businesses, and middle-class consumers across the globe, changes that will affect the lifestyles of all but the poorest. These realizations spawn a deep anxiety, which can easily lead to depression, apathy, or despair.

When your nonprofit focuses on loss of rainforests, or species extinctions, or soil erosion, or the rights of the poor, or lobbying for environmental policies, you know the challenges up close and personal. Sometimes you celebrate phenomenal successes; your efforts literally change the world. At other times you struggle through setbacks, limited funding, outright opposition by those who are making money through these injustices, and a populace that often doesn't seem to care either way. Overall, you recognize, the climate is changing, but humans aren't (yet).

The time-worn response, "If you can't beat 'em, join 'em," remains a temptation for many who watch the effects of capitalism and systems of global exploitation. How can we bring about change on such a large scale? The idealism that shines in the eyes of young people and activists can become dimmed by the magnitude of the problems before us.

A Big Idea That Does Something

The notion of ecological civilization speaks directly to this situation. It does not replace the environmental nonprofits, the marches, and the activism; in fact, it relies deeply on all of them. But it supplements them in a crucial way by holding the big picture—the biggest picture—before our eyes. Think of it as realism extended to the long-term. Studying civilizational change is not utopian; it doesn't say, "Don't worry, the perfect civilization is just around the corner." Instead, the mandate is to roll up one's sleeves and begin to lay the foundations for what will come after the unsustainable practices of the modern era end—whether they end in 5 years or 50, through gradual transition or violent collapse.

So, why does the prospect of civilizational change bring hope?

(1) *It offers a direction.* Reflecting on the possibility of an ecological civilization makes us think hard about how human

civilization will have to be organized for it to become genuinely sustainable. It helps us to render that goal more concrete and specific than it has ever been before. As our conceptions of what a sustainable and just society entails become clearer, they begin to offer more guidance for policies today. We have at least a rough roadmap of where we need to go, which means we have some idea of how to get there. Wandering aimlessly in the wilderness can induce hopelessness; but knowing your destination, no matter how big the mountains that stand between here and there, begets hope—we know where we need to go.

(2) *It grows out of a comprehensive vision.* The next hope is contained in the term itself. Ecological civilization formulates the long-term goal of humanity: to live together in a global civilization that can be sustained over the long run—a civilization where resources are shared and where the many do not suffer for the sins of the few. The comprehensive nature of the crisis we face can only be adequately addressed by an equally comprehensive set of solutions. As a vision for civilizational change, ecological civilization encourages a systematic strategy for systemic transformation. It's not simply the high stakes, but the never-ending laundry list of problems needing to be solved which many find overwhelming. By taking a comprehensive approach, the long and fragmented list of problems are reoriented through the lens of an integral civilizational system—meaning the countless separate social and environmental problems are brought together as one civilizational problem, which feels more manageable and instills a sense of hope.

(3) *It's already being realized here and now.* The previous hope, rooted in the possibility of a roadmap, begets a second one: activity. Many things need to be done now if we are to succeed in reaching the goal. It is human nature to feel better when we can make a difference; human despair comes when there is absolutely nothing for us to do. Some readers will remember a time when we believed that, if we recycled bottles

and cans, drove less (or drove a Prius), and thought globally while acting locally, then we would be saving the environment. Today we know these steps won't by themselves stave off climate change. At this stage we define success differently. It is no longer about avoiding climate effects or extinctions; they are already here. Instead, success is now about the progressive movement toward a new civilization, and the actions that take us in that direction are all valuable—whether we avert an economic collapse or begin rebuilding after it passes. As we saw in Chapter 7 above, the movement toward a new civilization is already upon us. An ecological civilization is beginning to emerge.

(4) *The goal is achievable.* The final ground for hope may seem strange, but it's perhaps the most profound of all: realism. Suppressing a frightening possibility for the future, trying not to think about it, often deepens anxiety. By contrast, bringing the fears out of the closet and looking them in the eyes brings a strange kind of freedom. Between today's "modern" civilization and a sustainable form of life on this planet lies a great chasm. The threat is real that we don't make it, that humanity does not respond in time. If we don't, it's likely that we will witness a collapse of many of today's social, economic, and political structures—of modern globalization as we know it. The consequences will be huge. There is something immensely freeing about looking that possible result in the face. When we know what we're facing, we know what we most want to avoid.

Against the backdrop of a sober assessment of the global situation, there remains the possibility that we might just make it. Knowing where we are now (realism), knowing where we are going (goal), knowing what we need to do (roadmap), doing it (activism), and knowing we may just make it (hope) is transformative. The journey to a sustainable civilization may be quick, with amazing technological breakthroughs, massive sharing of resources on the part of the powerful, and voluntary self-sacrifice on the part of the rest. Or we may first have to walk through the valley of the shadow of death. But either way, we—some of us—will get there. Thinking hard about the long-term outcome brings the deepest and the most realistic hope of all. It is the vision behind this book.

Creating a Framework, Fostering Collaboration

The previous chapter focused on examples of how the idea of ecological civilization is motivating new practices at the grassroots level. Now we want to step back and find out *why* it has spawned these actions. How is it different from environmentalism more generally, and from other "green" movements toward sustainability?

There is a strong pressure for policy people to be experts in one particular area, say food, water, or energy. We *do* need experts, and we *do* need to embrace sustainable practices in agriculture and economics, to support the development and use of renewable energy, to encourage lifestyles that lessen the amount of carbon dioxide and methane in the atmosphere. Many nonprofits are doing brilliant work in one of these (and many other) sectors. Nonprofits that do not do policy work often support grassroots innovations and movements, or they develop new and inspiring visions and narratives for the future. Relatively few focus on concrete policy needs that bridge across multiple sectors of society.

Think of a continuum from least to most comprehensive:

1. A specialized discussion of a single issue: saving leopards in India; safely disposing of nuclear waste;
2. Specialists in a sector, such as the energy sector;
3. Bi-sector discussion: how energy and transportation interrelate;
4. Discussion across sectors with regard to an issue, say, the dozens of specific steps that can be taken to reverse climate change;
5. The framework for putting together the various steps and analyses into a global plan of action;
6. A new story, paradigm, or worldview, such as integral thinking;

advocates claim that, if we have it clearly in our minds, we will solve climate change.

The notion of ecological civilization functions at the level of (5). It looks forward from the first four to the last and from the last back to the individual steps. It means looking at the issues holistically, seeking interconnections between sectors, and building on the connections across sectors that are most crucial for achieving long-term change.

Consider the relationship between *fostering collaboration* and *creating a framework*. Fostering and sustaining collaboration means growing networks into a movement, supporting them through organizational structures, maintaining communication, creating an "actionable constituency," and organizing conferences and public actions. Groups like 350.org, Greenpeace, the Sierra Club, and the World Wildlife Fund do this extremely well. Creating an overarching framework is different. It's not the same as fighting for carbon reduction or forest preservation, as essential as these are. In these pages we have shown how the sustainable civilization of the future must become the framework in light of which all facets of society can be redesigned. The overall magnet, as it were, is not specific environmental victories—of course we must win these as well!—but civilizational change itself; and all actions are assessed based on their role in helping to achieve this goal, as steps along the way. Only if you keep your eye on the whole forest will you know how to save the individual trees.

Holistic, systemic, long-term—these are important guides for describing, structuring, and carrying out environmental action. Tracing the interconnections all the way up to the civilizational level, one learns to understand, refine, test, and apply them. *This* is the necessary framework for sustaining collaborations across a whole range of sectors: food justice, agricultural reform, access to drinkable water, the empowerment of women and girls, social and individual lifestyle change. Think tanks working on specific policies, and organizations advocating for specific reforms, are the heart of environmental transformation. But long-term coordination requires integration across the sectors of society if, for example, we are to see religious traditions working together as allies or to build significant partnerships between governments, businesses, and NGOs.

Even multi-sector proposals are sometimes not enough. We think of the example of Paul Hawken's recent book, *Drawdown*, which calls itself "the most comprehensive plan ever proposed to reverse global warming."[3] The "Drawdown agenda" consists of 100 solutions to climate change, which he divides into sectors: energy, food, women and girls, buildings and cities, land use, transport, materials, and "coming attractions."[4] Make no mistake; Hawken does a great service by collecting initiatives from so many different sectors into one book. Yet we still need to ask: what place does climate change, and the dozens of factors that contribute to it, play within the broader movement of civilizational change? What will sustain the kind of hope that is required for the sacrifices that lie ahead? Beyond the list of tasks there is also a need for the kind of global thinking that connects these different areas *conceptually*. Increasingly, one discerns the urgency of higher-level integration across sectors, a need for thinking at a theoretically more sophisticated and higher level about how one can produce change.

"Backcasting"—How to Guide Action toward the Goal

The goal, then, is clarity about the long-term goal so that it can serve as a guide to action today. Some scholars are using the term *backcasting* to describe this process. Backcasting is "a reverse-forecasting technique which starts with a specific future outcome and then works backwards to the present conditions."[5] Since the 1970s it has served as a method for addressing major societal challenges. We can often identify an outcome that we want to achieve, perhaps 50 years in the future, but we are unsure of what steps we should take today to get there. In design research, people "propose a future event or situation and then work backward to construct a plausible causal chain leading from here to there."[6] What is typical is for people to try finding solutions by just building forward from the present toward the future. In backcasting, by contrast, we describe the desired outcome in as much detail as we can and, with these results in hand, derive "short-term planning and policy goals that might facilitate a long-term outcome."[7]

Question #8

The term "reverse engineering" is used in a similar way, although its connotations are somewhat different. Taken primarily from computer science, this concept means taking apart a completed product, usually hardware or software, to see how it works, so that one can put together pieces that have the same result. More broadly, reverse engineering is a way of moving from whole to part; one then puts parts together, differently than one would otherwise have done, in order to create the desired outcome. Taken as a metaphor, it suggests carefully analyzing—taking apart—the broader concept of ecological civilization in order to see how we can organize the parts of society now so that they are more likely to produce the outcome we seek.

Whatever the term, the key idea is the same: one begins with the future in order to guide actions in the present. It is necessary to trace and retrace the circle: *backcasting* from the outcome we seek—a sustainable society—in order to guide decision-making today, *forecasting* the outcomes of the today's policy decisions as steps toward that goal, which helps to further clarify the goal, which sharpens the backcasting, which further orients actions today.

This is ecological civilization in action. One works backwards from the idea of a different kind of civilization, one based on ecological principles across every sector. Backcasting sets the standards for environmental policymaking today; scholars and leaders can contrast what we are actually doing with what we would have to do. For example, current practices in the energy and transportation sectors accelerate climate disruption, which has become a threat to existing economic, social, and political systems. Investments in alternative energy sources and sustainable forms of transportation help us envision more concretely what a society fully based on ecological principles would look like. Backcasting once again, policy makers can set more precise goals and strategic priorities in the present.

This method is not being widely employed by governments, religious leaders, or, in many cases, by leaders of NGOs. Interestingly, it is common in the private sector. Businesses often calculate future patterns of supply and demand, estimating resource availability and consumer demand. They then make their investments based on their calculations of probable future states. Presumably it is these sorts of

155

calculations that have led Shell Oil to take the lead in developing hydrogen-powered cars. The Drawdown project, mentioned above, aggregates methods for slowing climate change across multiple sectors, which is an important step forward. Using backcasting, however, it becomes possible to ask: how we can integrate and prioritize the various steps in light of specific civilizational goals? What is the final product in terms of which the various ingredients should be put together?

Like Captain Picard of *Star Trek,* we, too, are set out on a journey "to boldly go where no one has gone before." The vision of ecological civilization needs to be flexible, since we are discovering for the first time what it requires. As we work backward from the end goal to inform actions in the present, these insights feed back into our understanding of the end goal itself—less like a straight line and more like a figure 8. As it unfolds, the process forces us to look at the truths of our situation: the increased inequality between the rich and the poor, and most of all the impossibility of unlimited growth on a finite planet. It creates a new map for a new destination. Just as a strategic plan for an organization is built from its mission and vision, a strategic plan for a nation—or a planet—must be guided by a unifying goal. Otherwise one gets what we currently see: a grab-bag of plans and policies at best, gridlock otherwise. Even in the nonprofit sector we have difficulties building longer-term collaborations across NGOs. Yet piecemeal efforts by themselves are not enough; they cannot create a critical mass across sectors sufficient for transformation at national and international levels. The sum of today's parts is a far cry from the new whole that we urgently need.

It is not difficult to imagine examples of integrative work of this kind.[8] They usually begin with broad-based policy innovations that respond to systemic global issues. Cape Town comes close to running out of water completely; how can other cities prepare for predicted water shortages in their own regions? The leaders of the world's religions meet once every four years to coordinate their social outreach; how can experts best communicate to them what is known about the climate crisis? The food crisis among the poor is affected by multiple factors locally, regionally, and globally; how can experts be brought together to produce more systemic and effective responses?

Question #8

This bridging function remains crucial. Many organizations work in specific sectors of the environmental crisis; fewer do multi-sector work. There are also many more "big vision" groups that talk about what the future as a whole should be, but without the attention to details that is required in order to guide actions in the present. Nonprofits built on careful backcasting play a crucial role in developing policy outlines and action plans that are responsive to the systemic challenges of the current global situation.

Partnerships in a Global Network

What are the minimum agreements needed to put the backcasting method to work? First, one has to agree on the goal. We believe the goal is a genuinely ecological civilization, and there are other ways to describe it, as we saw in Question Five. Howeverthe goal is named, such a civilization can be built only by people who think first in terms of organisms and ecologies, rather than primarily in terms of machines and individuals. We present the distinction using *pairs* of terms, because the rethinking involves both the units that make up the world and how they are related. If the drive for independent consumption by individuals has taken humanity to the edge of collapse, then a society built on ecological principles will have to start with *inter*dependence—moving cooperation rather than competition to the center. More efficient technologies are good. But until people are actually living for a shared good, they have not yet fully joined the movement of transformation.

In these pages we have focused on the long-term vision, a vision that goes beyond what many are doing today. We are surrounded by significant efforts at reform, and it is important that we support and assist them. Current reform efforts include virtually every sector of society: energy, transportation, trade, development, agriculture, education, urbanization, economics, politics, international relations—the list goes on. At this point, however, the goal has to be to *trans*form and not only to *re*form current structures. To climb Mount Everest, you have to reach Base Camp and then go beyond it to Camps 1, 2, 3, and 4, and then even beyond them. In the same way, we need to reach and then go beyond Environmentalism, beyond "Going Green," beyond Sustainable Development, beyond Materialism, beyond Capitalism, beyond Socialism, even beyond Eco-Justice.

Take the example of renewable energy. Many who support the use of renewable energies such as wind and solar do so with the intent of mitigating carbon dioxide emissions while maintaining current (and growing) levels of energy consumption. The term "sustainability" is often used in this sense to mean sustaining current lifestyles (particularly for those whose lifestyles are such that they want to continue them!). By contrast, while an ecological civilization may promote the use of renewable energy, it also entails *a far more fundamental shift* beyond our current systems of exploitation—of nature as well as of the marginalized—a shift in mindset that then affects lifestyles and levels of consumptions.

Now imagine experts across multiple sectors of society who think in terms of the longest-term goal of achieving an ecological civilization and backcast from that goal to their own fields. The goal affects their study of and recommendations regarding their own fields. It compels them to meet together to study interconnections among sectors, and indeed to establish working groups that use backcasting to guide multi-sector analysis. Partnerships of this sort are by their very nature transformational. These ongoing collaborations produce concrete plans for change, plans that are broader and longer term than the standard political cycle normally allows.

These examples help us to highlight the difference it makes to work on the basis of the organic or ecological model. Organizations and individuals who are participating in the ecological civilization movement have developed paradigms, models, and methods that challenge standard disciplinary assumptions. Because scholars such as Peter Brown at McGill University and Jon Erickson at the University of Vermont have moved beyond the "environmentally sensitive economics" that they were trained in and have become advocates of a genuinely "ecological economics." They are beginning to develop the principles and forms of trade that will contribute to the flourishing and stability of a *post*modern, ecological society, and the steps we will then have to take to get there.

Question #8

One also begins to see the kinds of new partnerships that are growing within the movement. Philosophers, historians, and religious leaders help initially to clarify the paradigms that will underlie a postmodern society. The more clear and compelling their visions are, the more helpful they will be for experts who use backcasting to guide them in their fields. Policy experts, private sector visionaries working for the common good, lawyers, and a few political leaders can have a huge influence in "operationalizing" the new policies and in giving them the political, legal, and financial support that they need. Like Kumsil Kang, Former Minister of Justice in South Korea, who is fighting for Earth jurisprudence, even giving legal rights to nature, we need visionary leaders to offer creative solutions inspired by the long-term goal of a world that works for all. Finally, heads of nonprofits and grassroots activists can already begin to put genuinely ecological principles into practice, "bringing ecological civilization down to earth" and "giving it a street address," as Eugene Shirley, the president of Pando Populus, describes the work of his organization.[9]

A framework called *network theory* has arisen in order to describe interactions of this type.[10] The process is not hierarchical; advances in any one area resonate through the whole network. Each contributor is a node in the network; the actions of one can permeate through the network to affect any of the others. Those who want to focus on the general goal, the mode of thinking that is needed to work toward ecological civilization, contribute to the overarching concepts. Think tanks play a continuing role in working through implications and scenarios. Grassroots leaders are already experimenting with putting these ideas into practice, and their knowledge is indispensable for scholars and political leaders. It does not take complete agreement on any one term or theory in order to make a positive contribution to this movement.

What results is a network of mutual encouragement across the ecological civilization movement. People working in a particular field see needs and opportunities that other organizations are not aware of and ask for assistance from those working in other areas; in other cases they can act directly. It is also important to bring ideas from ecological civilization think tanks to organizations that are in a position to implement them. For example, when Nancy Minte, the president of Uncommon Good, organizes immigrant farmers in south Pomona to grow vegetables in the midst of a so-called food desert, she needs lawyers and city officials to help make changes in zoning laws so that the farmers can legally sell their produce to people in their neighborhoods. On a more global level, the water crisis in Cape Town cannot be solved by hydrologists alone; a huge range of expertise, not only from South Africa, but also around the world, is required to conceive and implement the necessary transformations in thought and practice as we reconceive urbanization in an age of drought.

Network-based cooperation for the common good can be implemented at local, regional, national, and international levels. Complex networks are characterized by diverse approaches and emphases; if the nodes lack diversity, the network is doomed to failure. It takes the longer-term focus on ecological civilization to retain the radically pluralistic perspective that is necessary to sustain cooperation within the movement. In the early stages more informal structures will suffice, but over time they will have to grow and become more established. Remember that, at the present rate of change, within a few years the situation will outgrow whatever structures we have been able to develop for this particular stage.

That is why we speak of a global movement and network: it depends on cooperative interactions from the grassroots level to the international level. The network stretches from your own carbon footprint and composting for your garden, to microeconomic loans for women in African villages, to vast financial assistance for nations where populations are dying because of climate disruptions that they did not cause. And it involves, finally, a different world- and life-view that affects us all the way down to the innermost motivations that guide our actions. Some in the movement are calling this "ecological spirituality."

We are watching how hope is reborn as people begin to see roadmaps for the transition from present to future. Look at a roadmap of England, and you will conclude that "all roads lead to London." Look backwards from the goal of a sustainable world, and you will see the roads from each sector of human civilization converging on that single goal, like spokes toward the center of a wheel. The journey may be torturous, and the costs in terms of human lives and species extinctions may go beyond what we can comprehend. But the map with converging roads is clear. Don't let anyone tell you that "we just don't know what to do." For each sector, at least a rough description of the steps is already available; and the more we walk them, the clearer the next steps become. It's time to get on our feet and start walking.

Realistic Hope

In these pages we have looked unflinchingly at the reality of a planet in crisis. Human activity has raised the heat-trapping gases in our environment to the highest level in 800,000 years. The consequences of that mistake are now globally visible. So is the future. As a scientifically advanced species, we have extremely detailed information on what the effects will be on every continent. We are on a train careening out of control down a mountain slope, and we have only a short time to apply the brakes before the train runs off the tracks. Unfortunately, we know a lot more about what the crash will look like than one might wish to know. Denying the reality of our situation does not change it.

Percy Bysshe Shelley has authored what may be the most powerful poem on civilizational change ever written. With a few unforgettable images, he captures the contingency of human civilization and the pride of the kings and technocrats who think that their empires and multi-national corporations will never pass:

> I met a traveller from an antique land,
> Who said—"Two vast and trunkless legs of stone
> Stand in the desert. . . . Near them, on the sand,
> Half sunk a shattered visage lies, whose frown,
> And wrinkled lip, and sneer of cold command,
> Tell that its sculptor well those passions read
> Which yet survive, stamped on these lifeless things,
> The hand that mocked them, and the heart that fed;

And on the pedestal, these words appear:
My name is Ozymandias, King of Kings;
Look on my Works, ye Mighty, and despair!
Nothing beside remains. Round the decay
Of that colossal Wreck, boundless and bare
The lone and level sands stretch far away."

Modern civilization has been the first civilization in human history to expand to cover the entire globe; its achievements are legion. But in its hubris it has become Ozymandias. Deserts have arisen where civilizations once thrived, their ruins surrounded by lone and level sands.

Genuine hope springs up when illusion dies. Shelley's harsh image brings an essential reminder: no civilization is above the possibility of collapse. Here the knowledge of the scientist and the historian converge.

But the situation is not hopeless. The actions that the experts are calling for are correct . . . and urgent. NGOs around the world, courageous individuals, religious communities, and a few governments and businesses are taking concrete, even sacrificial steps to live out the necessary changes. We urge you to join them.

At this point, however, attitudes *about* what we are all doing, and assessments of our odds, are all over the map. Some environmentalists and activists are optimistic that, if we put our backs into it, we can avert disaster. Others, looking at the data and the current global response, conclude that it is already too late. Attitudes matter. Living without hope incapacitates one for action, but living with naïve hope sets one up for disabling disappointment.

This book has been about a third possibility—an approach that guides thought and action, but that also speaks to this crucial question of attitude. To begin to comprehend the global situation as one of civilizational change leads to a very different kind of hope: a realistic hope, a long-term hope. This planet will still nurture life a few centuries from now, and in almost all the scenarios humans will still be among its inhabitants. But by all accounts, the path between here and there will be rocky. How rough it will be depends on which bookmakers you consult. Some put odds on a few bumps but not a crash; something like our

Question #8

present civilization and lifestyles will come out the other side, albeit with a few loses and a number of improvements. (Maybe we can still fly on our vacations, but just with electric airplanes.) By contrast, the Seizing an Alternative movement reads the data as saying that human society will be fundamentally changed. A temperature increase around the globe of 3 degrees Celsius and above—that is, hotter by 5.4 degrees Fahrenheit or more, which is no longer unlikely—leaves virtually nothing untouched.

A realistic hope has two foundations. The first is now: let's roll up our sleeves, change our lifestyles, strengthen sustainable community, and gain skills for self-reliance, which will radically lower our carbon footprint. Let's share this knowledge, motivate others, and work to build national and international movements. It's already happening around the planet; humanity is in motion. If we put our minds (and our shoulders) to it, humanity is still able to turn the ship around.

And if we don't? Hope for the future remains. If our current civilization creates new deserts, unfarmable fields, undrinkable water, unbreathable air, then its infrastructure will crumble, and it too will pass. If the survivors create another unsustainable civilization, its fate will be the same. Only a sustainable civilization will survive and thrive over the long term—after all, that's what sustainable means. There is only one long-term solution, then: a genuinely ecological civilization. To recognize this, and to begin to take steps to establish it, is the foundation for realistic hope.

So let's get to work. Every positive step we take today has a dual purpose: it decreases the severity of the crash that lies ahead, and it begins the transition toward a sustainable society. Whether it is the second generation (your grandchildren) or the seventh generation, our descendants will need a new set of skills and a new way of thinking. It gives us incredible hope to know that we are rediscovering those skills, altering our lifestyles, and formulating the worldview that they will need if they are to live in an ecological way on this planet.

Endnotes

1 John Cobb raised this question back in 1971 with his book, *Is It Too Late?: A Theology of Ecology*. While a bit more optimistic in the 70s,

the lack of significant change in the decades that followed has left Cobb confident in collapse, but not without hope. As such, his recent work has focused less on preventing collapse and more on building the foundation for a better future after collapse. We, the authors, believe that regardless of whether or not collapse is avoidable, our path is the same—we need to take steps toward an building an ecological civilization (see Chapters 6 & 7).

2   That's about 75 billion tons of living things. See "How many living things are there?" USCB Science Line, http://scienceline.ucsb.edu/getkey.php?key=1388.

3   Paul Hawken, *Drawdown: The Most Comprehensive Plan Ever Proposed to Reverse Global Warming* (New York: Penguin, 2017).

4   See, for example, "Food Sector Summary" at the Drawdown website, https://www.drawdown.org/solutions/food.

5   "Backcasting," BusinessDictionary, http://www.businessdictionary.com/definition/backcasting.html.

6   "Backcasting," Design Research Techniques, http://designresearchtechniques.com/casestudies/backcasting/.

7   "Backcasting," Design Research Techniques.

8   See EcoCiv.org for more details and further examples.

9   See PandoPopulus.org.

10 See for example Bruno Latour, *Reassembling the Social: An Introduction to Actor-Network-Theory* (Oxford: Oxford University Press, 2005). In science and technology studies Latour, Michel Callon, and John Law are important contributors to actor-based network theories. Network theory has also been developed in Computer Science; see Andrey Kurenkov, "A 'Brief' History of Neural Nets and Deep Learning," http://www.andreykurenkov.com/writing/ai/a-brief-history-of-neural-nets-and-deep-learning/.

www.ingramcontent.com/pod-product-compliance
Lightning Source LLC
LaVergne TN
LVHW050608191224
799473LV00035B/915